Respectful Rehabilitation

MASONRY

How to Care for Old and Historic Brick and Stone

Respectful Rehabilitation

MASONRY

How to Care for Old and Historic Brick and Stone

Mark London

The Preservation Press

The Preservation Press
National Trust for Historic Preservation
1785 Massachusetts Avenue, N.W.
Washington, D.C. 20036

The National Trust for Historic Preservation in the United States is the only national, private nonprofit organization chartered by Congress to encourage public participation in the preservation of sites, buildings and objects significant in American history and culture. Support is provided by membership dues, endowment funds, contributions and grants from federal agencies, including the U.S. Department of the Interior, under provisions of the National Historic Preservation Act of 1966. The opinions expressed in this publication do not necessarily reflect the views or policies of the Interior Department. For information about membership, write to the Trust at the above address.

Library of Congress Cataloging in Publication Data
London, Mark
Masonry: how to care for old and historic brick and stone.

(Respectful rehabilitation series)
Bibliography: p.
Includes index.
1. Masonry. 2. Building, Brick. 3. Building, Stone.
4. Historic buildings — Conservation and restoration.
I. Title. II. Series.

TH1199.L66 1988 693′.1 86-25270
ISBN 0-89133-125-5

Printed in the United States of America

92 91 90 89 88 5 4 3 2 1

Edited by Diane Maddex, director, and Janet Walker, managing editor, The Preservation Press.

Designed by Meadows & Wiser, Washington, D.C.
Composed in Galliard and Geometrica by General Typographers, Inc., Washington, D.C.
Printed by the John D. Lucas Printing Company, Baltimore, Md.

Cover: Repointing a townhouse on Capitol Hill, Washington, D.C. The work is being performed by Landmark Restoration Inc. with a high-lime, tinted mortar appropriate for this historic building at 420 10th Street, N.E. Cover photograph by Carol Highsmith

Contents

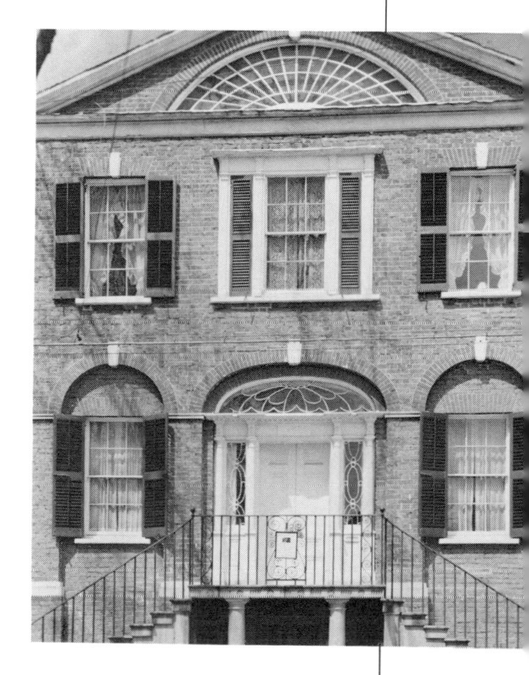

FOREWORD

As president of the National Trust for Historic Preservation — charged with the care of a score of historic properties and the responsibility to help all Americans preserve their own special places — I know all too well the demands of maintaining buildings that are centuries or even a few decades old. Masonry is a subject of particular concern to us, because so many of our properties are brick and stone.

At Lyndhurst (1838, 1864–65), Jay Gould's castle in Tarrytown, N.Y., craftsmen have replaced deteriorated marble on the Gothic Revival facade with a custom-made substitute — cast stone mixed with marble dust from the deteriorated pieces to match the original color and texture. A master stonemason helped us reconstruct a dry-laid retaining wall at Oatlands, near Leesburg, Va. The wall on the early 19th-century estate was carefully rebuilt by taking the wall apart and then placing each stone back to ensure a tight fit. Drayton Hall (1738–42), near Charleston, S.C., an outstanding example of Georgian architecture and the Trust's oldest property, has needed almost continual repointing of its 18th-century brick. The mortar was soft to begin with, a situation not helped by poor repointing over the years with too hard a mortar.

Reconstruction in progress on the dry stone wall at Oatlands. (Dean Korpan)

All of the National Trust's experience underscores the fact that caring for old buildings means much more than just wanting to save them. It means that we need to maintain and rehabilitate these places with the respect that their age and history deserve.

In publishing this book, the National Trust shares with you the most up-to-date information and techniques about how to care for your own masonry buildings. With this clear and authoritative advice, rehabilitating with respect should be easier than it ever has been.

J. JACKSON WALTER, President
National Trust for Historic Preservation

Opposite: Main entry at Lyndhurst, whose exterior stone veneer is a local marble no longer available. (Jack E. Boucher, HABS)

A·D
1897

Preface

I live in a brick and stone building and work in another. I spend a lot of time staring at masonry walls, looking at the subtle differences among them and considering how they were constructed.

Stone walls: some in shades of taupe, others in mushroom or smoky gray. Some made of stones uniformly cut, polished and carefully aligned; others of roughly hewn rock set in a crazywork quilt, seemingly without order until they reveal subtle patterns that even the masons may not have known they were creating.

Brick walls: on one building, the bricks are glass-smooth, rust red with wafer-thin mortar joints flush with the face of the brick. On the next, the bricks are salmon orange with mortar joints recessed just enough to create a crisp dark shadow across the surface of the mortar so that part of the joint is almost white, while the shaded part is a warm charcoal. On another building, bricks stand upright across the tops of windows, like soldiers with arms locked together, to help support the weight of the wall above them.

Window details. The soldier course of bricks across the lintel, bonding pattern, color and surface appearance of the brick all help to give this building character. (Dinu Bumbaru)

Stored up in an old wall are the efforts of dozens of craftsmen. There were workers who made the bricks by digging the clay out of the ground, taking it in horse-drawn wagons to the plant, tamping it into wooden molds, laying the bricks to dry in the hot sun, carefully stacking them in the kiln and making the hot wood fire. There were those who made the mortar by gathering and grinding oyster shells and who burned their hands while slaking the lime. And there were the masons who built the wall, straight and true.

The craftsmen of past generations built to last, and the brick and stone buildings they created can last for centuries, if they are not mistreated. Masonry is one of the most durable building materials, but at the same time it is particularly vulnerable to the two most common repair activities: cleaning and repointing. One day of ill-considered sandblasting can

Opposite: Detail of a sculpted sandstone entrance to a home in Montreal. (Dinu Bumbaru)

Stereoscopic view of a workman turning bricks to dry at a New York State brick company. (Library of Congress)

sentence a brick wall that has survived for generations to a short life of a few years. A few hours of crudely slathered repointing work can obscure the subtle relationship of form, spacing and depth between joint and stone that the early designer and mason so carefully created.

This book has been written for people responsible for the care of old and historic buildings of brick and stone — owners of houses and commercial buildings, administrators of schools and churches, and anyone else who wants to maintain or rehabilitate masonry structures respectfully. It shows how to evaluate, understand and direct rehabilitation work on brick and stone. Architects, contractors and other professionals who deal with old buildings on a daily basis will also find this a useful overview of the subject. The book's purpose is to bridge the gap between very general home renovation books that have a few pages on masonry but may not properly consider historic buildings and technical works that are often of little practical use to the average property owner.

Although the National Park Service cooperated in the preparation of this book, and some parts of it are based on National Park Service material, the views expressed here do not necessarily represent official Interior Department or National Park Service policies, and some techniques discussed may not generally be approved by the National Park Service for work on historic buildings. Before planning rehabilitation that may be eligible for federal tax credits, consult with the National Park Service or a state historic preservation officer.

Mark London

Acknowledgments

This book is the result of an international collaboration among the Preservation Press of the National Trust for Historic Preservation, the National Park Service of the U.S. Department of the Interior and the Heritage Montreal Foundation. Although not specifically referenced in the text, much of the material is based on publications of these organizations that are here revised, condensed, excerpted, expanded and brought up-to-date.

In particular, several publications of the National Park Service have been most helpful. Preservation Brief no. 1, *The Cleaning and Waterproof Coating of Masonry Buildings* by Robert C. Mack, AIA, and no. 6, *Dangers of Abrasive Cleaning to Historic Buildings* by Anne E. Grimmer, served as the basis for the chapter on cleaning brick and stone. *A Glossary of Historic Masonry Deterioration Problems and Preservation Treatments* by Anne Grimmer was used in writing much of the chapter on repair. Preservation Brief no. 2, *Repointing Mortar Joints in Historic Brick Buildings* by Robert Mack, de Teel Patterson Tiller and James S. Askins, served as the basis for the chapter on repointing. *Moisture Problems in Historic Masonry Walls: Diagnosis and Treatment* by Baird M. Smith, AIA, guided the text of the chapter on moisture problems, and certain elements of the diagnosis methodology Smith proposed in that book are reflected in that chapter.

Material from the Heritage Montreal Foundation's *Traditional Masonry* by Dinu Bumbaru and me, published in cooperation with the Quebec Ministry of Cultural Affairs and the Heritage Canada Foundation, has found its way into every chapter. That publication and preparation of the present one benefited from generous financial assistance to Heritage Montreal from the Samuel and Saidye Bronfman Family Foundation. Heritage Montreal's *Carrying Out a Renovation Project* by Robert Paradis was used as a basis for the Introduction's section on hiring a contractor.

Sculpted sandstone element, adding a whimsical touch. (Pietra Dura, Inc.)

Some of the details of the surface repair techniques described in the chapter on repairing masonry are taken from the New York Landmarks Conservancy's pamphlet *The Maintenance and Repair of Architectural Sandstone* by Michael F. Lynch and William J. Higgins. The original includes more detailed instructions and names of suppliers.

In addition, *Masonry Conservation and Cleaning,* a compilation prepared by Keith Blades for the 1984 conference of the Association for Preservation Technology, was an invaluable resource in assembling this book. Details on all of these publications, including ordering information, can be found under Further Reading.

I would especially like to thank the following people who generously shared their knowledge and advice: Norman Weiss of Columbia University; H. Ward Jandl and Anne E. Grimmer of the Technical Preservation Services branch, National Park Service; Keith Blades of the Canadian Ministry of Public Works; Dinu Bumbaru of the Heritage Montreal Foundation, who also provided the illustrations; Baird M. Smith, AIA, Washington, D.C.; and Martin Weaver and Herb Stovel of the Heritage Canada Foundation. Of course, none of these people is responsible for any deficiencies in the book.

This project would never have gotten under way without the initiative and support of Diane Maddex, director of the Preservation Press, who was a pleasure to work with throughout the preparation, and of Phyllis Lambert and Jean-Claude Marsan, past and present presidents of Heritage Montreal. The manuscript was carefully edited and produced by Janet Walker, managing editor of the Preservation Press, with assistance from Paul C. Wolman, Michelle LaLumia and Pamela Haag.

This book is dedicated to my wife, Linda Thompson, who generously gave both moral and practical support to this work, and to all other lovers of old brick and stone buildings.

Opposite: Theophilus Conrad House (1893–95), Louisville, Ky., a Richardsonian-style mansion built of rough-hewn limestone. (HABS)

2023
R. G. DeSANTIS
VETRONE
468-3710

INTRODUCTION

There is no doubt that when people rehabilitate they want to make their buildings better. But a walk through almost any old neighborhood shows that many "improvements" have quite the opposite effect on a building's character and durability.

Why cover a century-old limestone facade with artificial stone? Why take a sandstone wall whose lime mortar has come out in only a few places and repoint the entire wall with hard cement mortar that cracks the stone? Why sandblast brick, destroying its protective surface, and then cover the wall with a synthetic coating that may trap moisture and increase the probability that the brick will crumble?

Why do careful consumers blithely sign contracts for thousands of dollars of renovations that end up doing tens of thousands of dollars of damage to their homes while destroying their architectural character?

People usually choose what seems to be the cheapest and easiest solution to a problem. They are often simply not aware of the implications of what they are doing. Buildings are complex, so it is easy for property owners to feel overwhelmed; seeking security, they put themselves in the hands of the contractor "who has been doing things this way for decades." No wonder they are dazzled by salespeople with briefcases full of charts and figures "proving" that their products are the only solution for a building's problems.

Ironically, with the best of intentions, people often spend far more than is necessary to solve a problem. Consider instead the preservation axiom: the less changed, the better. The least expensive and technically best solution is also almost always the one that best respects the character of the building; it is the one that requires the least change. In the situations mentioned above, the appropriate solutions described in this book would almost certainly have cost less and would not have damaged the buildings.

Top: Permastone-covered townhouse next to an original brick facade. (A. Pierce Bounds)

Above: Sandblasted brick doorway. (Mary Means)

Opposite: Townhouses in Philadelphia, built of brick and faced with brownstone ashlar. (HABS)

Getting Help

In general, masonry work is not the part of a rehabilitation project that inexperienced homeowners should take on themselves. They should direct their energies to less risky jobs such as interior painting and finishing. The few masonry repairs that may be undertaken successfully by homeowners include simple water washing of brick or stone and small repairs such as replacing a few bricks on a stairway or pavers in a sidewalk. Beyond these tasks, homeowners should seek the assistance of an architect or contractor.

Perhaps the main reason for inappropriate treatment of many masonry problems is the easy accessibility of commercial products and services. Large companies with healthy marketing budgets can afford to promote their wares with aggressive salespeople, big ads in newspapers, magazines and the Yellow Pages, and large displays in hardware stores. But finding the appropriate treatment usually takes more than a phone call. It may require a certain effort to understand the problem, to choose an appropriate treatment and to find just the right masonry contractor. And it will mean deciding on the exact roles expected of each professional hired. Above all, those interested in respectful rehabilitation should consider calling in a restoration architect or other reliable independent expert for everything but the simplest problem.

Hiring an Architect

Because masons and masonry contractors may not be able to place the immediate problem in the context of the whole building, they cannot plan an integrated, coordinated solution if more than one part of the structure is involved. The mason may treat the symptom without having analyzed and dealt with the fundamental cause of the problem. Contractors and companies also may be biased in favor of products they sell or services they offer.

An independent building expert, most often an architect, should be put in charge, not only of major rehabilitation projects, but also of any masonry problem that is at all out of the ordinary. The architect should be responsible for surveying the existing building, analyzing the problem, outlining possible courses of action with cost estimates, preparing plans and specifications, calling for bids and supervising the work.

Even for a small and relatively simple job, a brief consultation with an architect — based on an hourly rate — might help avoid serious problems. In some cases, it may be adequate to have the architect look at the building and suggest a plan of action without preparing full contract documents or supervising the contractor who would carry out the work.

Seeking advice. Even a short consultation with a restoration architect or specialist in historic preservation can mean the difference between success or failure in repairing a masonry structure. (Sid Latham)

Choosing an architect

In choosing an architect, remember that very few have special training or experience in rehabilitation or restoration. Begin with a local preservation organization, historical society, architects' association or city preservation office and ask if persons there can recommend architects who have been responsible for recent rehabilitation projects and especially masonry work. Seek estimates from two or three architects, check their references and discuss the job with them before making a choice.

An architect's fee usually varies between 5 and 15 percent of the cost of the work depending on the scale and complexity of the project. An hourly rate may range from $50 to $125. This expense should be returned many times over by avoiding unnecessary work, mistakes and inefficiency.

Plans and specifications

The architect should prepare complete plans that describe the project. In addition to the plans and, if necessary, annotated photographs, which visually describe the work to be done, written specifications are usually prepared, describing the work in detail, including materials to be used and how they will be installed. For a simple job, this might require just a few pages. For a building rehabilitation, a complete "spec" (list of specifications) is required; this is organized by trade, each section detailing the scope of work, the materials and the execution. Materials will include special stone or brick, which must be ordered well in advance.

The more complete the contract documents, the better protected an owner is against misunderstandings and unforeseen costs. In addition, good plans and specifications that clearly outline the scope of work are essential in making a fair comparison among the bids of potential contractors.

Hiring a Mason or Contractor

If a job seems related only to the brick or stone of the building, a mason or masonry contractor can be hired directly. If the job is more complex and involves many trades, it is best to hire a general contractor to run the whole project.

General contractors take full responsibility for completing

an entire job; they, in turn, usually hire subcontractors (masons, carpenters, plumbers, electricians) to execute various parts of the job. The general contractor coordinates all the work and charges about 15 percent of the total cost of the job. Amateurs' attempts to manage jobs themselves can easily result in poor coordination that increases costs far more than the contractor's fee.

Contractors also are usually in a better position than a property owner to find the best subcontractors, to negotiate a good price and to get the best work.

Choosing a contractor

The choice of a contractor should be made with care. It is best to hire local, established contractors who depend on the satisfaction of their customers for repeat business.

Traditionally, the term "mason" referred only to people working with stone, while those dealing with brick were called bricklayers. Now, the term is used broadly to refer to people who work with stone, brick and concrete block. Many masons and masonry contractors deal mainly with new construction and have little knowledge of the special problems of old brick and stone. In fact, many of today's masons have never worked with stone at all.

Here are some of the key steps in choosing a mason or contractor:

- Get names suggested by the project architect, the municipality, a local historical organization, owners of recently rehabilitated buildings, friends or neighbors.

- Get at least three written, itemized bids. Costs of preparing the bid are usually covered by revenue from contracts actually received. Do not necessarily take the lowest bid. Be suspicious if the low bid bears little relation to the other bids; make sure the contractor really understood the scope of work.

- Ensure that the contractor is a member of the local contractors' association and inquire if this provides the owner any special guarantee. Ask about the contractor's insurance.

- Ask the contractor for the names of several recent clients and the addresses of the projects. Check how reliable the contractor was in respecting the contract (quality of work, budget, schedule) and whether any charges for extra work added later were reasonable. Check whether any complaints have been registered at the Better Business Bureau.

Costs

In analyzing your estimates, remember that work improperly done results in a loss in the building's value and the cost of repairing the damage can be many times the entire cost of the project. Cutting corners can represent false economy.

Because the cost of materials generally represents a relatively small part of the overall costs, it is worthwhile to specify the best possible materials, even if it means custom ordering special bricks or stones to match the rest of the masonry.

Costs in a project depend not only on the masonry work itself, but also on other site conditions and ongoing work. For example, repointing a wall will cost more if the damaged part is accessible only by scaffolding. However, if the masonry repair is part of a larger rehabilitation project, the cost of renting scaffolding can be apportioned to other work, such as the repair of windows, woodwork and the roof.

Remember also that the quality of the work depends to a great extent on the preparation. A well-prepared job may cost more but will last much longer.

Insurance

Building owners should be certain that the contractor and subcontractors have full liability and fire insurance on the building. If the owners are contracting for the work themselves, they should arrange for this insurance. Insurance coverage of the building should be regularly increased as rehabilitation progresses and the value of the building grows. For minor work, such as small home repairs, a temporary rider might be added to the homeowners policy to cover the extra risk during rehabilitation work.

The contract

Most contracts for small or medium-size jobs are given on the basis of a fixed price, plus unit prices for specific items or work, such as a unit price per window sill or per square feet of repointing. For larger or more complex jobs other arrangements may be made, such as paying the contractor all the direct costs plus a percentage for overhead and profit or by hiring a separate construction management firm. The project architect should be consulted on such procedures.

Standard contract forms for engaging contractors should be available from local architects' or contractors' associations.

The contract should include the following:

- Name and address of the owner and that of the mason or contractor
- Detailed description of the work to be done or a reference to the plans and specifications, which then become part of the contract
- Overall value of the contract, a breakdown of costs and when payment will be made
- Dates when the work will begin and end; the bonus the contractor is to receive if work is completed early and the penalty if it is completed late — usually a per diem amount
- Statement requiring that the contractor conform to all federal and local codes
- Type and duration of bonds
- Civil liability of each party
- Unit costs for additional work that may be required, for instance, the cost per linear foot to repoint or per brick to be replaced
- Statement requiring that the contractor clean up the site at the end of each day, each week and thoroughly upon completion of the work
- Indication whether the owner or the contractor is responsible for obtaining the necessary permits

Each part of the contract including appendixes should be signed. Remember that anything not spelled out in the contract probably will not get done.

Payments are generally made as the work progresses (every month or 15 days). Before each payment the contractor should supply a release demonstrating that all workers and suppliers have been paid for the previous month. An amount of 10 or 15 percent should be retained (the hold-back) until 30 days after the work has been completed to ensure that it is satisfactory and all subcontractors and suppliers have been paid.

For a small job, for example, a few thousand dollars to get the building cleaned or repointed, a simpler payment schedule can be used, such as 35 percent when the work is half completed, 50 percent on completion and a hold-back of 15 percent to be paid 30 days later.

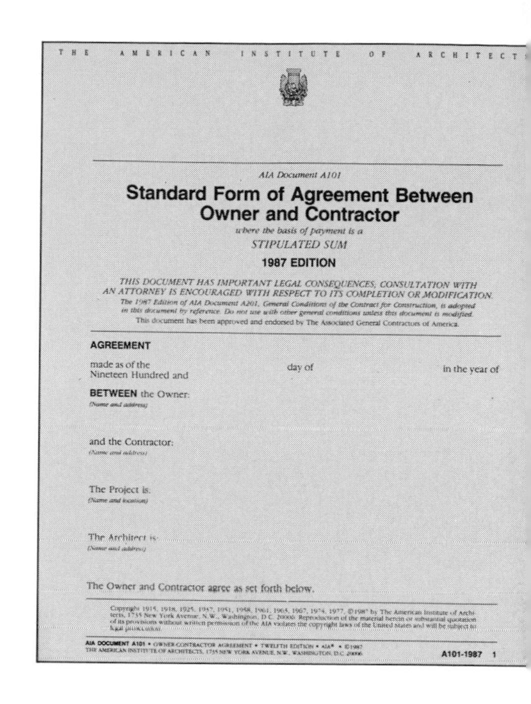

THE AMERICAN INSTITUTE OF ARCHITECTS

AIA Document A101

Standard Form of Agreement Between Owner and Contractor

where the basis of payment is a

STIPULATED SUM

1987 EDITION

THIS DOCUMENT HAS IMPORTANT LEGAL CONSEQUENCES; CONSULTATION WITH AN ATTORNEY IS ENCOURAGED WITH RESPECT TO ITS COMPLETION OR MODIFICATION.

The 1987 Edition of AIA Document A201, General Conditions of the Contract for Construction, is adopted in this document by reference. Do not use with other general conditions unless this document is modified.

This document has been approved and endorsed by The Associated General Contractors of America.

AGREEMENT

made as of the day of in the year of Nineteen Hundred and

BETWEEN the Owner:
(Name and address)

and the Contractor:
(Name and address)

The Project is:
(Name and location)

The Architect is:
(Name and address)

The Owner and Contractor agree as set forth below.

Copyright 1915, 1918, 1925, 1937, 1951, 1958, 1961, 1963, 1967, 1974, 1977, ©1987 by The American Institute of Architects, 1735 New York Avenue, N.W., Washington, D.C. 20006. Reproduction of the material herein or substantial quotation of its provisions without written permission of the AIA violates the copyright laws of the United States and will be subject to legal prosecution.

AIA DOCUMENT A101 • OWNER-CONTRACTOR AGREEMENT • TWELFTH EDITION • AIA® • ©1987
THE AMERICAN INSTITUTE OF ARCHITECTS, 1735 NEW YORK AVENUE, N.W., WASHINGTON, D.C. 20006

A101-1987 1

Standard owner-contractor form of the American Institute of Architects.

Taking Care of Priorities

The first priority in repairing masonry is to deal with the primary causes of deterioration. Before stones cracked by settlement are replaced or repaired, the foundations themselves must be stabilized. Before bricks spalled because of excessive moisture are replaced, the source of water infiltration must be eliminated. This may involve major work on other parts of the building such as new site drainage, underpinning of foundations or a new roof. Once this preliminary work is completed, repair of the masonry itself can begin.

Here, too, one should start with major structural work such as reinforcing sagging arches or rebuilding badly deteriorated walls before undertaking minor repairs, such as repointing or cleaning.

Sometimes it may be necessary to do urgent temporary repairs, perhaps while waiting until sufficient funds are available to go ahead with a complete rehabilitation. These may include dealing with dangerous conditions, such as shoring up a wall likely to collapse. Temporary repairs may also be required to prevent further deterioration, such as installing a plastic or metal covering on badly deteriorated areas to prevent water from continuing to penetrate and doing even more damage.

Discussing the day's work. On any rehabilitation project, lines of communication and authority must be clear. (Lamont Green)

Once the architect, general contractor or masonry contractor has been hired, set up a chain of command and stick to it. Workers need to receive instructions from one source, or chaos reigns; if a problem arises, discuss it with the contractor, not the worker.

If during the course of the work charges for extras are required, establish the cost in advance and have a written description (change order) signed by the contractor.

Make sure that the municipal building inspector checks the building at appropriate stages of the work to avoid having to demolish an area to look at new work that has been hidden. For instance, have the inspector check foundation insulation before earth is backfilled or the air or vapor barrier before the wall is bricked up.

If there are serious problems with the work of the mason or contractor, bring in an independent inspector, who can later serve as a witness, if necessary. Take lots of photographs; record weather conditions and any other pertinent information.

Reviewing design plans prepared by an architect for a downtown Madison, Ind., businessman. (James L. Ballard)

Most masonry repairs, if done correctly, should last many years. In approaching a masonry repair project, good intentions will not guarantee a good job. What is needed is an understanding of masonry materials and what harms them, coupled with a cautious, critical approach on everyone's part — property owner, architect, contractor and subcontractor. This book highlights some basic concerns and can help prevent the worst errors. Even a few hours of reading and additional research will make an owner or designer familiar with building methods and the wide range of good materials, elements and techniques available for rehabilitation.

Preserving a Building's Character

An 18th-century farmhouse in New England, a Civil War–era store in Chicago, a turn-of-the-century office building in San Francisco, an Art Deco hotel in Miami Beach — each building has a special character that deserves respect. Each design — ornate or simple — combines local traditions and materials with the new technologies and changing styles that have swept the continent every few decades during the past two centuries. The smallest detail can contribute to the look of a building; subtle nuances reflect an architect's careful design decision or craftsman's skill handed down from generation to generation.

Today, the harmony that came from the limited availability of materials and the strictly defined building traditions of our predecessors is gone. With a virtually limitless choice of materials, techniques and styles, we have swept away the old traditions. Structurally, and aesthetically, anything goes.

As liberating as our new technical dexterities and eclectic aesthetics are, when they spread to our diminishing stock of old buildings, they can lead to insensitive repairs and changes that upset or destroy the integrity of a building's original design. The ill-considered juxtaposition of styles and materials in a building — such as adding rustic Mediterranean-style pointing or colonial windows to a Victorian home — transforms old buildings into structural smorgasbords belonging to no particular date, style or tradition. Even seemingly minor changes can downgrade the appearance of buildings or even of whole neighborhoods because of the cumulative effect of many small transformations.

This chapter looks at the principles involved in choosing maintenance and rehabilitation techniques that respect a building's character.

Ocean Drive in Miami Beach's Art Deco District. (HABS)

Opposite: Fox Theatre (1929), Atlanta, a renowned movie palace brought back to life in the late 1970s. (HABS)

The Less Changed, the Better

It is important to preserve original design features of all old buildings, not just those of recognized historical significance. It may be tempting to make changes, but creating a new composition as coherent as the original is difficult. In addition, a building's authentic, original materials themselves have an intrinsic historical value. In the past few decades, these attitudes have become widely accepted and are reflected in the marketplace; just as an old armoire is worth more if it still has its original hardware and patina, so the market value of today's buildings reflects the thoughtfulness with which they have been preserved.

The basic principle to follow in preserving and rehabilitating an old structure is to respect the character and design of the building and its features. *The goals are to retain architectural elements, avoiding as much as possible changing or replacing them, and to ensure that alterations or additions respect the original design.* The greater the architectural and historical value of the building, the more diligent one should be in preserving the original features and the more reluctant one should be to change anything.

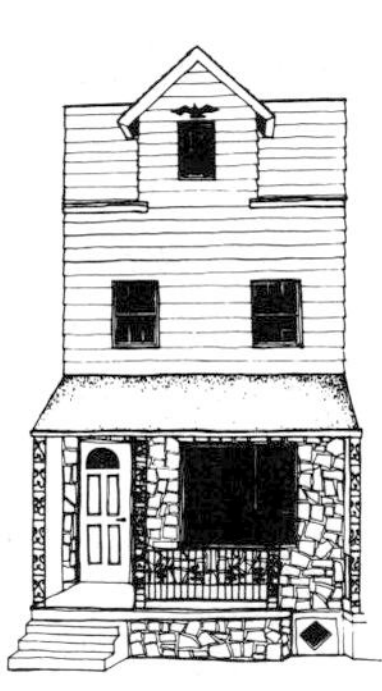

An old house as originally constructed (top), appropriately modified (center) and inappropriately altered with new windows, siding and permastone (above). (Old Allentown Preservation Association)

Right: New Jersey farmhouse with awkward rear addition and poor foundation repointing. (HABS)

Frank Lloyd Wright's Frederick C. Robie House (1908–09), Chicago, an architectural landmark in the Prairie Style. (Cervin Robinson)

Defining a Building's Significance

The significance of a building, whether derived from its design or its association with historic events or people, is not limited to the oldest and most remarkable few typically labeled "historic." Many of the most charming streets and attractive neighborhoods so vital to our heritage are made up of fine buildings that may not be individually designated or listed in any official landmarks register. The advice and information in this book, although essential for the most historic of our buildings, apply equally to all old buildings, especially those more than 40 years old.

A wide range of factors, including its age, affect the role a building plays in its environment:

Architectural design. What is the quality of a building? Is it unique in some way? Is it the work of a well-known architect or builder?

Historical significance. Did a historic event occur at the site? Did important cultural, religious, social, political, economic or industrial activities take place there? Is the structure associated with an important individual or organization?

The Alamo, San Antonio, Tex., whose historical significance is unquestioned. (Zintgraff Photographers)

Top: Wheeling, W. Va., townhouses in a historic neighborhood. (Jack E. Boucher, HABS)

Above: Quaker School, Lincoln, Va., part of a historic district. (Ashton Nichols)

Context. What is the building's relation to its surroundings? What is its role in establishing the character of its area? Is it a visual landmark?

Integrity. Does it still have most of its original features? Do previous changes or additions undermine its overall character? Can inappropriate modifications be corrected?

It is not easy for a property owner to answer these questions, and the advice of an architect, historian or preservation consultant will probably be necessary.

The greater the architectural and historical value of a building, individually or as part of a group, the more important it is that rehabilitation respect the building's character. Structures of exceptional value demand meticulous restoration. Even if a building is not exceptionally important, it should not be unduly altered when it has an interesting character and contributes to the quality of a neighborhood.

A building is a rich accumulation of layers reflecting many uses and changes over the years. It is generally preferable to keep those changes that have, over time, become part of the overall character of the building, rather than strip away layers in an attempt to reconstruct a long-lost original state that may not be documented. If the original is truly intact underneath the layers, the action may be justified; what is far worse is imprecise reconstruction of elements long gone. The question of which modifications are significant and merit preservation is a difficult one. For buildings of great historical significance, the owner should make this judgment after seeking an outside, objective opinion. If the renovations are not already subject to a review by a local design board or historical commission, consider submitting the project to an independent expert, perhaps at a local university's school of architecture.

In a brick or stone building, the masonry is obviously fundamental to its overall character. However, some parts of the building may be less critical, such as a poorly designed recent addition that is incongruous or, sometimes, secondary elements such as the backs of row houses, which might be modified without harming the overall character of the building. Before touching any element, the following questions should be asked: Is the element original? Does it have notable qualities of design or craftsmanship? And, in general, how significant is it in defining the building's character? If the answers are not obvious, seek the advice of an independent expert.

Row house in Baltimore, before and after rehabilitation. The integrity of each design element has been retained. (Doug Barber)

Basic Approaches

To preserve a building's character, an owner should seek to put the building in good condition so it can serve contemporary needs while changing as little as possible of the building's materials and appearance.

It is better to maintain than to repair. If a building is properly maintained in good condition, major repairs may never become necessary, although small repairs are a part of ongoing maintenance. Regular inspection and immediate correction of small defects will usually forestall costly major repairs and replacement.

Damaged elements should be repaired rather than replaced. It is often possible to patch, piece in, splice, consolidate or otherwise reinforce or rebuild damaged elements. If this is not enough, replacement of portions of extensively deteriorated or missing parts or features by carefully matching surviving parts may be necessary. Although using the same kind of material is preferred, substitute materials sometimes may be acceptable if the appearance matches the original. Features should not simply be removed and not replaced in an attempt to create a "modern" appearance or to avoid maintenance costs.

Day-Taylor House (1856), Hartford, Conn. This Italian-style villa underwent a major restoration in the late 1970s. (Roger V. Dollarhide)

If repair is not possible, the replacement should match the original as closely as possible. If deterioration or damage to materials precludes repair, replacing features with like material is recommended. Replacements should match the original features in design, texture and detail. Substitute materials may also be acceptable if they convey the visual appearance of the surviving parts of the masonry feature and are physically and chemically compatible. If an entire feature is missing, the original can be re-created through historical research and by examining old photographs or investigating the building itself. A replica of the feature can be produced based on the results of this research. Or, the feature can be replaced with a modified design that is compatible in size, scale, materials and style.

Err on the side of conservatism. Ask an independent expert to evaluate proposed changes to an old building before proceeding with work. If it is not absolutely clear that changes are necessary, do not make them. If changes do become necessary in order to ensure a building's continued use, avoid radical alterations or additions and those that obscure or destroy significant spaces, materials, features and finishes. Also, try to make it as easy as possible to return the building to its original condition in the future by documenting the building's condition before and during changes. If possible, store any decorative elements that have been removed. Alterations and additions should be avoided if at all possible in buildings of great significance; if built, additions should be clearly differentiated from the historic structure.

Rehabilitation Standards and Guidelines

The Department of the Interior has developed standards and guidelines that have become widely used by the federal government as well as by a variety of other organizations and preservationists. Although developed for historic buildings, the principles are universal and provide sound advice for anyone who wants to maintain a building's character.

The Standards

The standards are used as the official criteria by which work on historic properties listed in the National Register of Historic Places is evaluated and eligibility for federal tax credits is certified. Such credits are available principally to income-producing businesses.

In the standards, "rehabilitation" is defined as "the process of returning a property to a state of utility, through repair or alteration, which makes possible an efficient contemporary use while preserving those portions and features of the property which are significant to its historical, architectural and cultural values." This assumes that at least some repair or alteration of the building must take place to provide for an efficient contemporary use; however, the secretary of the interior's standards state that these repairs and alterations must not damage or destroy the materials and features — including their finishes — that are important in defining the building's historic character.

"Restoration" is defined in the standards as "accurately recovering the form and details of a property and its setting as it appeared at a particular period of time by means of the removal of later work or by the replacement of missing earlier work." This implies that the only changes to be made would be those that return the property to an earlier look.

The standards for rehabilitation are just one set of the Interior Department's overall standards for historic preservation projects, which also include sections on other types or degrees of treatments. Although the rehabilitation standards apply to most situations, in specialized circumstances specific standards for acquisition, protection, stabilization, preservation, restoration or reconstruction should be used.

The Secretary of the Interior's

Standards for Rehabilitation

1. Every reasonable effort shall be made to provide a compatible use for a property which requires minimal alteration of the building, structure or site and its environment, or to use a property for its originally intended purpose.

2. The distinguishing original qualities or character of a building, structure or site and its environment shall not be destroyed. The removal or alteration of any historic material or distinctive architectural features should be avoided when possible.

3. All buildings, structures and sites shall be recognized as products of their own time. Alterations that have no historical basis and which seek to create an earlier appearance shall be discouraged.

4. Changes which may have taken place in the course of time are evidence of the history and development of a building, structure, or site and its environment. These changes may have acquired significance in their own right, and this significance shall be recognized and respected.

5. Distinctive stylistic features or examples of skilled craftsmanship which characterize a building, structure or site shall be treated with sensitivity.

Cobblestone Schoolhouse (1819), Childs, N.Y., a distinctive regional building technique. (Jack E. Boucher, HABS)

Second level of the former Petty Building, a native sandstone structure in Hot Springs, S.D. (James L. Ballard)

6. Deteriorated architectural features shall be repaired rather than replaced, wherever possible. In the event replacement is necessary, the new material should match the material being replaced in composition, design, color, texture and other visual qualities. Repair or replacement of missing architectural features should be based on accurate duplications of features substantiated by historic, physical or pictorial evidence rather than on conjectural designs or the availability of different architectural elements from other buildings or structures.

7. The surface cleaning of structures shall be undertaken with the gentlest means possible. Sandblasting and other cleaning methods that will damage the historic building materials shall not be undertaken.

8. Every reasonable effort shall be made to protect and preserve archeological resources affected by or adjacent to any project.

9. Contemporary design for alterations and additions to existing properties shall not be discouraged when such alterations and additions do not destroy significant historical, architectural or cultural material and such design is compatible with the size, scale, color, material and character of the property, neighborhood or environment.

10. Wherever possible, new additions or alterations to structures shall be done in such a manner that if such additions or alterations were to be removed in the future, the essential form and integrity of the structure would be unimpaired.

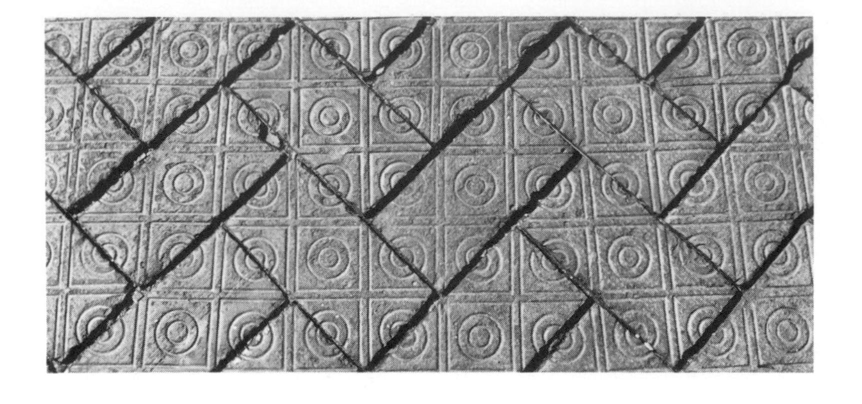

Decorative brick sidewalk, Galesburg, Ill. The brick was manufactured at a local brickworks. (Clark Schoettle)

THE GUIDELINES

The Department of the Interior has also developed a set of guidelines that provide general design and technical recommendations to help property owners, designers and building managers meet the standards. The guidelines are general and are not meant to give case-specific advice or address exceptions or rare instances.

According to the guidelines, owners of historic properties have two priorities: to preserve the significant architectural materials and features and to make possible an efficient contemporary use. Possible actions are listed in order, starting from the least intrusive (evaluation and maintenance) and moving to the greatest and most problematic change (new construction):

- Identify, retain and preserve, clarifying which architectural materials and features are important in defining the historic character
- Protect and maintain
- Repair
- Replace, only under certain well-defined circumstances

The guidelines also include sections on design of missing features, alterations and additions to historic buildings.

Included here are all the guidelines for rehabilitation that deal with masonry in general and brick and stone in particular. Approaches, treatments and techniques consistent with the standards are listed under the Recommended column in the table. Actions that could adversely affect a building's historical character are listed under the Not Recommended column.

Guidelines for Rehabilitating

Historic Masonry Buildings

Brick, stone, terra cotta, concrete, adobe, stucco and mortar

Masonry features (such as brick cornices and door pediments, stone window architraves, terra cotta brackets and railings) as well as masonry surfaces (modeling, tooling, bonding patterns, joint size and color) may be important in defining the historic character of the building. It should be noted that while masonry is among the most durable of historic building materials, it is also the most susceptible to damage by improper maintenance or repair techniques and by harsh or abrasive cleaning methods. Most preservation advice for masonry thus focuses on such concerns as cleaning and the process of repointing. For specific guidance on this subject, consult Preservation Brief nos. 1, 2, 5, 6 and 7.

Preservation

Recommended	Not Recommended
Identifying, retaining and preserving masonry features that are important in defining the overall historical character of the building such as walls, brackets, railing, cornices, window architraves, door pediments, steps and columns; and joint and unit size, tooling and bonding patterns, coatings and color.	Removing or radically changing masonry features which are important in defining the overall historical character of the building so that, as a result, the character is diminished. Replacing or rebuilding a major portion of exterior masonry walls that could be repaired so that, as a result, the building is no longer historic and is essentially new construction.

Recommended	Not Recommended
Protecting and maintaining masonry by providing proper drainage so that water does not stand on flat, horizontal surfaces or accumulate in curved decorative features.	Failing to evaluate and treat the various causes of mortar joint deterioration such as leaking roofs or gutters, differential settlement of the building, capillary action or extreme weather exposure.
Evaluating the overall condition of the masonry to determine whether more than protection and maintenance are required, that is, if repairs to the masonry features will be necessary.	Failing to undertake adequate measures to assure the preservation of masonry features.

Cleaning

Recommended	Not Recommended
Cleaning masonry only when necessary to halt deterioration or remove heavy soiling.	Cleaning masonry surfaces when they are not heavily soiled to create a new appearance, thus needlessly introducing chemicals or moisture into historic materials.
Carrying out masonry surface cleaning tests after it has been determined that such cleaning is necessary. Tests should be observed over a sufficient period of time so that both the immediate effects and the long-range effects are known to enable selection of the gentlest method possible.	Cleaning masonry surfaces without testing or without sufficient time for the testing results to be of value.
Cleaning masonry surfaces with the gentlest method possible, such as using low-pressure water and	Sandblasting brick or stone surfaces using dry or wet grit or other abrasives. These methods of cleaning

Recommended	Not Recommended
detergents and natural bristle brushes.	permanently erode the surface of the material and accelerate deterioration.
	Using a cleaning method that involves water or liquid chemical solutions when there is any possibility of freezing temperatures.
	Cleaning with chemical products that will damage masonry, such as using acid on limestone or marble or leaving chemicals on masonry surfaces.
	Applying high-pressure water cleaning methods that will damage historic masonry and the mortar joints.

Painting

Recommended	Not Recommended
Inspecting painted masonry surfaces to determine whether repainting is necessary.	Removing paint that is firmly adhering to, and thus protecting, masonry surfaces.
Removing damaged or deteriorated paint only to the next sound layer using the gentlest method possible (e.g., handscraping) prior to repainting.	Using methods of removing paint which are destructive to masonry, such as sandblasting, application of caustic solutions or high-pressure water blasting.
	Applying paint or other coatings such as stucco to masonry that has been historically unpainted or uncoated to create a new appearance.
	Removing paint from historically painted masonry.

St. Louis, Mo., townhouses in the Soulard Historic District, unified by material, color, scale and height. (Robert Pettus, HABS)

Recommended	Not Recommended
	Radically changing the type of paint or coating or its color.
Applying compatible paint coating systems following proper surface preparation.	Failing to follow manufacturers' product and application instructions when repainting masonry.
Repainting with colors that are historically appropriate to the building and district.	Using new paint colors that are inappropriate to the historic building and district.

Repointing

Recommended	Not Recommended
Repairing masonry walls and other masonry features by repointing the mortar joints where there is evidence of deterioration such as disintegrating mortar, cracks in mortar joints, loose bricks, damp walls or damaged plasterwork.	Removing nondeteriorated mortar from sound joints, then repointing the entire building to achieve a uniform appearance.
Removing deteriorated mortar by carefully hand-raking the joints to avoid damaging the masonry.	Using electric saws and hammers rather than hand tools to remove deteriorated mortar from joints prior to repointing.

Recommended	Not Recommended
Duplicating old mortar in strength, composition, color and texture.	Repointing with mortar of high portland cement content (unless it is the content of the historic mortar). This can often create a bond that is stronger than the historic material and can cause damage as a result of the differing coefficient of expansion and the differing porosity of the material and the mortar.
	Repointing with a synthetic caulking compound.
	Using a "scrub" coating technique to repoint instead of traditional repointing methods.
Duplicating old mortar joints in width and in joint profile.	Changing the width or joint profile when repointing.

Repairing

Recommended	Not Recommended
Repairing masonry features by patching, piecing in or consolidating the masonry using recognized preservation methods. Repair may also include the limited replacement in kind — or with compatible substitute material — of those extensively deteriorated or missing parts of masonry features when there are surviving prototypes such as terra cotta brackets or stone balusters.	Replacing an entire masonry feature such as a cornice or balustrade when repair of the masonry and limited replacement of deteriorated or missing parts are appropriate. Using a substitute material for the replacement part that does not convey the visual appearance of the surviving parts of the masonry feature or that is physically or chemically incompatible.

Recommended	Not Recommended
Applying new or nonhistoric surface treatments such as water-repellent coatings to masonry only after repointing and only if masonry repairs have failed to arrest water penetration problems.	Applying waterproof, water-repellent or nonhistoric coatings such as stucco to masonry as a substitute for repointing and masonry repairs. Coatings are frequently unnecessary, expensive and may change the appearance of historic masonry as well as accelerate its deterioration.

REPLACEMENT

Recommended	Not Recommended
Replacing in kind an entire masonry feature that is too deteriorated to repair—if the overall form and detailing are still evident—using the physical evidence to guide the new work. Examples can include large sections of a wall, a cornice, balustrade, column or stairway. If using the same kind of material is not feasible, then a compatible substitute material may be considered.	Removing a masonry feature that is unrepairable and not replacing it, or replacing it with a new feature that does not convey the same visual appearance.

Jacksonville, Ore. An arch has been filled in over the doorway, interrupting the harmony of the original arcade. (Jack E. Boucher, HABS)

Stucco

Recommended	Not Recommended
Repairing stucco by removing the damaged material and patching with new stucco that duplicates the old in strength, composition, color and texture.	Removing sound stucco or repairing with new stucco that is stronger than the historic material or does not convey the same visual appearance.

Adobe

Recommended	Not Recommended
Using mud plaster as a surface coating over unfired, unstabilized adobe because the mud plaster will bond to the adobe.	Applying cement stucco to unfired, unstabilized adobe. Because the cement stucco will not bond properly, moisture can become entrapped between materials, resulting in accelerated deterioration of the adobe.

Missing Historic Features

The following work represents particularly complex technical or design aspects of rehabilitation projects and should be considered only after the preservation concerns listed above have been addressed.

Recommended	Not Recommended
Designing and installing a new masonry feature such as steps or a door pediment when the historic feature is completely missing. It may be an accurate restoration using historical, pictorial and physical documentation or be a new design that is compatible with the size, scale, material and color of the historic building.	Creating a false historical appearance because the replaced masonry feature is based on insufficient historical, pictorial and physical documentation. Introducing a new masonry feature that is incompatible in size, scale, material and color.

ARBLE CO.

IDENTIFYING KINDS OF BRICK AND STONE

There is no single best way to clean or repair masonry. The many varieties of stone and brick, the many types of deterioration and the many possible solutions require an understanding of the material and its problems before deciding on a solution. With the information in this chapter, and the aid of an expert as needed, it is possible to identify most types of brick or stone in a building. This will help in diagnosing problems and selecting an appropriate cleaning or repair technique.

Before focusing on the specific masonry ailments, examine the material itself. What is it? How was it made? What are its inherent strengths and weaknesses? Learn more about materials than just their general names. For example, the properties of sandstone vary widely: some sandstones, because of their low carbonate content, can be cleaned with acidic cleaners, whereas others must be cleaned with another technique. The properties of brick also vary: a well-fired, hard face brick can stand up to the rigors of weather, but a soft brick made for an internal wall that was exposed after the building next door was demolished could deteriorate rapidly and might have to be covered or replaced.

In any given area, experienced people in the construction and rehabilitation fields are usually able to help identify local stone and brick. But a restoration architect or the geology department at a local university may be needed to identify or even to lab-test unusual masonry brought in from another area. In addition, stone identification charts can help identify masonry types through a series of questions about color, texture, hardness and other characteristics. And samples of stones obtained from a local supplier can be compared with the stone in a building.

The following brief summary of the history, manufacture and basic physical and chemical properties of various types of

Squared limestone rubble wall. The masonry is more than 200 years old. (Dinu Bumbaru)

Opposite: Shipping area of a Vermont quarry around the turn of the century. (Library of Congress)

Top: Stratford Hall (c. 1725), Westmoreland County, Va., the home of the Lees. It was built of brick with massive brick chimneys. (NTHP)

Above: Rubble foundation. (© Richard Pieper)

stone and brick commonly used in North America can help an owner identify a building's masonry and the problems to which it is susceptible.

MASONRY USE IN NORTH AMERICA

An abundance of trees and a shortage of skilled labor led early European settlers in the Northeast to turn to wood for most building. Fieldstone and, later, crudely made brick were used primarily for foundations and chimneys. By the 17th century, the use of brick, popular in England and Holland, became widespread in the English colonies. The larger houses and public buildings of many New England towns were built of brick, while the use of stone persisted in French colonies and adobe was used in the American Southwest.

The growth of urban centers in the 18th century increased and concentrated the market for masonry, resulting in the establishment of stone quarries and plants to make brick using local clays. In the 19th century growing concerns about fire safety plus improved technologies and changing tastes led to even greater use of masonry. Cut stone was often used only for cornerstones (quoins) and to frame window and door openings in rubble walls. Later, cut stone was used for all of the main facades of buildings and then for entire buildings.

By the end of the 19th century, improved transportation encouraged the use of stone and brick from other localities and even from overseas. This new availability of a variety of masonry materials permitted full expression of late Victorian exuberance in architecture — a rich profusion of colors and textures of brick and stone, often in the same building.

With the spreading use in the late 19th century of steel and concrete for structural systems, stone and brick were relegated to exterior cladding. Terra cotta, a hard, usually hollow, molded ceramic material, also became popular, first for small decorative elements on brick facades and then for interior walls and floors. Glazed architectural terra cotta provided an inexpensive alternative to carved stone, and many steel- and concrete-framed buildings from the 1880s to the 1930s bear facades made entirely of terra cotta.

Concrete also came to be used as a facing material, and many 20th-century "stone" buildings are really concrete imitating stone. Early standard concrete blocks often had surfaces that mimicked the form of stone.

With the rise of modern architecture and industrialized building techniques in the 1930s to 1950s, brick and particularly stone were used less frequently as an exterior material for large public buildings, although brick continued to be used for houses and other small buildings. Many quarries closed or adapted to crushed-stone production for highway construction. In the 1980s the use of stone has been revived in new architecture, and many quarries have reopened to serve this need and the growing rehabilitation market.

Top: Morse-Libby Mansion (1863), Portland, Maine, an example of late Victorian experimentation with materials. (SPNEA)

Center and above: Terra cotta facade and artificial stone. (Dinu Bumbaru)

STONE

The term "stone" refers to a building material; in its natural environment it is called rock. There are three basic types of rock, and the term "rock cycle" is a simplified way of describing how the three types are formed geologically.

Igneous rocks are created by the cooling and crystallization of molten material (magma) in the earth. They form the cores of continents and mountain ranges and tend to be quite homogeneous and hard. A common igneous rock is granite.

Sedimentary rocks are formed by a more complex process. When weather and water action erode igneous and other rocks, the gravel, sand and silt that are formed wash into rivers, lakes and oceans and are deposited in layers on the sea or lake beds, along with the shells of marine organisms and minerals precipitated out of sea water. These particles are then cemented together by one or more binding materials into sedimentary rocks, a process called lithification. Sedimentary rocks cover most of the surface of the continents and are characterized by the presence of layers that are not necessarily of identical composition and that often contain fine layers (strata) of clay or other substances. The commonest sedimentary rocks are limestone and sandstone.

Metamorphic rocks are created when extreme pressure and heat in the earth re-form igneous and sedimentary rocks into a crystalline mass. Metamorphosed rocks are harder and denser; they are physically changed but have the same chemical composition as the rock from which they came. This category includes marble, transformed from limestone; slate, from shale; and gneiss, from granite.

The rock cycle is completed when, as a result of volcanic action, metamorphic and other rocks become molten again. In reality, the process is not quite so cyclical or clear cut; rocks are often in transitional states between any two of the three types.

Rock suitable as building stone has been quarried in most parts of North America, although half the U.S. production of building stone comes from Georgia, Indiana, Vermont and New Hampshire alone. The middle of the North American continent, once a vast inland sea contained by the Rocky Mountains to the west, the Appalachian Mountains to the east and the Canadian Shield to the north, is now made up largely of sedimentary rock from the Paleozoic age, 200 to 600 million years ago. Toward the center of the former sea area, in states such as Indiana and Missouri, are limestones formed from seashells. Closer to the ancient shores, in New York, Ohio and Pennsylvania, are sandstones. These rocks are generally solid and uniform enough to make excellent building stones. Along the Atlantic and Gulf coasts are more recent sedimentary formations (less than 60 million years old) that are generally not hard or uniform enough to use for construction.

A greater variety of rock is found in the mountainous areas. The Appalachians and the Canadian Shield, among the oldest geological formations on the earth, are largely Precambrian and more than 600 million years old. Here, rock layers have worn away, exposing ancient crystalline formations of granite, slate and marble. The Appalachians also have pockets, called Triassic basins, 180 to 230 million years old, of sedimentary rocks such as sandstone.

Making Stone

In the early days of North American settlement, the transformation of rock into stone consisted of gathering rocks found lying in the fields, often ones left by glaciers thousands of years earlier. These were simply carried to the construction site and used as they were, except that seriously offending projections would be knocked off. Rocks were also taken from sedimentary formations in river banks where the rock cracked readily into roughly rectangular shapes. Later, demand for stone brought about quarrying and dressing, with the degree of finishing based largely on how the stone was to be used.

Indiana limestone quarry around 1930. (Lake County Museum, Curt Teich Postcard Collection)

Top and center: Sawing stone with a diamond-toothed saw and chiseling stone columns. (Lake County Museum, Curt Teich Postcard Collection)

Above: Stonecutter's tools. (© Michael Devonshire)

Quarrying

At the earliest quarries, rock was hacked from exposed cliffs, canyons or other outcroppings. Later, rock was quarried from within the ground. Rock found near the surface tended to be weathered and to have small fissures that made it less durable as building stone. As quarrying techniques improved, this poorer rock was used only for rubble walls or was removed by blasting.

Because there is an abundance of good building stone in virtually all parts of the continent, early quarries were located close to where they were needed, generally near cities. Only in the latter part of the 19th century did transportation improve enough to allow stone to be moved economically far from local areas.

Quarrying continued to use ancient manual methods well into the second half of the 19th century. Sedimentary rocks were more often quarried because they were layered and large blocks could be split off easily. To split the rocks, quarriers drilled holes and drove in iron wedges with sledge hammers. Sometimes workers hammered in wooden wedges. When soaked, the wooden wedges expanded, causing the stone to split. In the 19th century quarriers filled holes with gunpowder, which was detonated with a burning fuse. More recently, holes were plugged with dynamite, which was detonated electrically. The shock of explosives causes hairline fissures in building stone, however. This then allows water to infiltrate, where it can erode, freeze and crack the stone. To improve the quality of stone at the end of the 19th century, quarriers introduced mechanized techniques using pneumatic shears, saws and hammers. Modern high-speed diamond drills and saws make the quarrying of harder and more durable stones such as granite more economical; granite now accounts for half the production of building stone in the United States.

After quarrying, large blocks of quarried stone were broken or cut to about their final size at the quarry to ease handling and reduce shipping costs. Originally done with wedges, this is now accomplished by sawing.

Dressing

In a rough rubble wall, a more or less rectangular stone could be used as shipped from the quarry. For walls of coursed rubble or squared stone (ashlar), the quarried stones first had to be squared (bankered up) into more regular rectangular form. Working with one straight edge of stone, a cutter used compasses, straight edges and an experienced eye to lay out the other three edges; then hammers and chisels, mallets and points, axes, picks and other tools were used to square off the stone. The exposed face might be finished to look like a rough, natural outcropping (rock faced) or made flat. In the latter case, the squaring would sometimes leave a chiseled or "drafted" margin of 1 to 2½ inches along each edge of the face.

The faces of the stone to be exposed were often further leveled using a variety of tools. These included the bush hammer (a grid of pyramidal points that created a stippled effect), tooth chisels and patent hammers (thin blades clamped together, which produced patterns of regular lines). Stones were then either polished or left with their distinctive surface tooling. Often, stones with several patterns were used on one building. Surface tooling is particularly susceptible to deterioration caused by weathering and inappropriate cleaning techniques.

The stones of column capitals, keystones and springing stones of arches, window and door lintels, pilasters and other parts of buildings were often carved. Softer stones such as sandstone and marble are easier to carve and thus were frequently used for highly decorative work, but because they are soft they are more subject to erosion.

Properties and Characteristics of Stone

A great variation exists in the way stone looks and in how well it works in construction. Color, hardness, texture, strength and resistance to weathering vary not only from one kind of stone to another but also between two pieces of the same type of stone from different quarries and even from stratum to stratum within a quarry. Understanding the characteristics of stone will assist in understanding why it is deteriorating and how it can be protected.

The following are the chief traits to look for in analyzing stone. Most are applicable to brick as well. If there are serious problems or major work is planned, it is worth determining the properties of the stone in detail, particularly if replacement stone is contemplated.

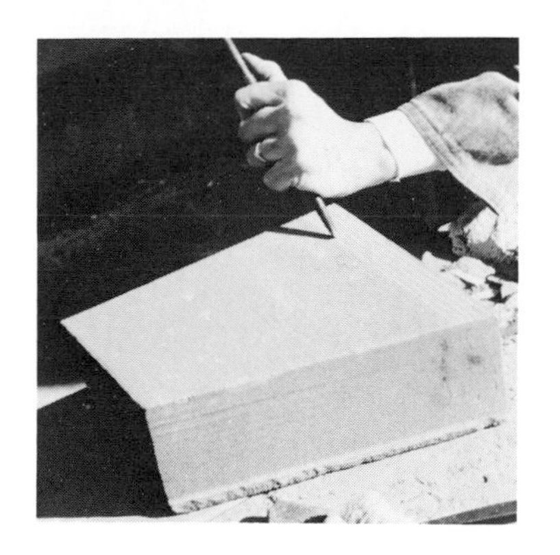

Top and center: Dressing stone. (© Michael Devonshire)

Above: Dressed stone used in construction. (Pierce Pearmain)

Moisture resistance

Most stone deterioration is caused by water-related problems. Moisture and frost action can make stone flake, powder, crumble and spall. Therefore, the most important factor affecting durability is the masonry's breathability as represented by its porosity and permeability (see Finding and Treating Moisture Problems for more detail).

Porosity is the ratio of pores (small voids) to the total volume. The ratio is low in igneous and metamorphic rocks, generally higher with sedimentary rocks. The size of pores is also important: the larger the pores, the more resistant the stone is likely to be to salt and frost damage.

Permeability is the ease with which liquids and gases flow through the stone. Systems of largely interconnected voids, or capillaries, run through stone, and their size and structure affect permeability. Narrower voids exert the strongest capillary action, which may transport liquid the farthest.

Acid and pollution resistance

Knowing how resistant a stone is to acids indicates how well it will stand up to pollutants such as acid rain and to cleaning with acid. Stones made up mainly of carbonate minerals such as limestone, marble and calcareous sandstones are particularly susceptible. Those based on silica such as siliceous sandstones are generally resistant. Polished stones with low acid resistance will lose their polish in a short time when exposed to acid rain. A simple way to test a stone's acid resistance is to drip a little hydrochloric (muriatic) acid on the surface. If bubbles form, the stone is not resistant to acids. More complete tests to evaluate the resistance of stones are being developed that involve immersion of the stone in a solution of dilute sulfuric acid.

Strength

The design of stone buildings reflects the fact that stone is strong in compression and weak in tension (pulling) and torsion (twisting). For example, small window or door openings can be spanned with a large stone lintel across the top. But longer spans can be made only by avoiding stone's weakness in tension — hence the arch, which is designed so that each stone is in compression. Again, igneous and metamorphic rocks tend to be the strongest, sedimentary rocks weaker, especially if they have high porosity. With modern construction, however, stone is usually used merely as a facing on a steel or concrete structure, and its strength is less important.

Density and hardness

Builders before the 20th century believed that stone with a density above 140 pounds per cubic foot would weather better. They did not have the technological skills to work stone of densities higher than about 170. We know now that even within this density range, porous and fractured stone can weather badly. And, since the 1920s, modern quarrying equipment has made denser and more resistant stones available in construction.

Hardness or toughness is how easily stone can be worked—that is, its resistance to impact and abrasion. Hardness is usually related to the density and strength of the rock and depends on its mineralogical composition. Igneous rocks tend to be the hardest and therefore are rarely sculpted. However, hardness is not always a function of density; some marbles are very dense but are relatively easily carved.

Friability is the susceptibility of stone to crumble naturally or to be broken easily. Weakly bound limestones are especially friable.

Color

The color of a stone is determined by its chemical composition. Generally, uniform colors, free of flaws or stains resulting from impurities, are favored for construction. However, varied colors and patterns are prized in decorative stones such as marble. In some stones, colors alter over time. Color can also change if the stone loses its polish. With sedimentary rocks, if the sediment was exposed regularly to the air when forming, as would have been the case at the edges of a receding inland sea, iron content would oxidize to make the rock reddish; otherwise, iron in rock would tend to remain gray or bluish. Color is not normally a factor in the durability of stone. But, for example, a reddish tone indicating the presence of iron oxides might be a useful warning against using the wrong kind of cleaning technique.

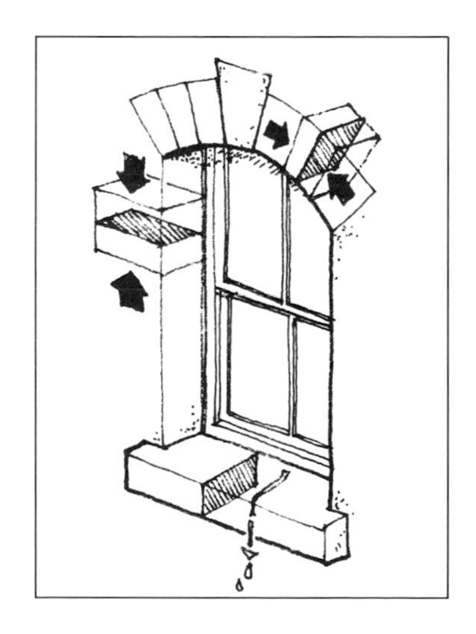

Bedding planes, laid perpendicular to loads of a wall and arch. In a sill, the plane should be laid so that only the edges are exposed. The sill should slope slightly to drain water.

Bedding planes

Sedimentary rocks are essentially laminated materials. The direction of the layers is called the rock's bedding plane. Because layers of weaker material such as clay are often sand-

Top: Granite (foreground), usually speckled in appearance, and limestone (background). (Dinu Bumbaru)

Above: Marble showing visible vein pattern. (Dinu Bumbaru)

wiched between bands of purer rock, stones cleave (break or split) most easily along the bedding plane and are strongest perpendicular to the plane.

Stones should be laid so that they are held together like a pack of cards, with their bedding planes perpendicular to compression loads. Also, stones are least likely to come apart if only the edges of the layers are exposed to the weather. For a normal stone in a load-bearing wall, this means that the bedding plane of most stones should be horizontal; in arches, bedding planes should be along the radius lines of the arch. However, for stones whose top horizontal surface is exposed, such as sills, cornices and copings on tops of walls, the stones should be turned so that the edges of the layers are upwards and the bedding plane follows the drainage of the water.

Often, for reasons of convenience, economy or lack of knowledge, stones were laid with the bedding planes in the wrong direction; the worst example was face bedding, or placing the bedding plane parallel to the building face. As a result, many buildings and graveyards are full of stones coming apart in layers.

Types of Stone

The four most common stones used in construction are limestone, sandstone, granite and marble.

Limestone

Limestones are sedimentary rocks consisting mainly of calcite (calcium carbonate) and/or dolomite (calcium magnesium carbonate). They are usually formed from fragments of shell, coral and other marine organisms. Some finely grained and compact limestones — for example, oolite — are created from chemical precipitates. Sometimes called greystone, limestones were popular for building because they combined relatively easy workability with good weather resistance. Sulfur oxides in today's acid rain converts limestone to friable gypsum, however.

Sandstone

Sandstones usually can be identified by their coarse, granular, sandy texture. They are sedimentary rocks that consist of consolidated sand grains (mainly quartz and feldspar) ce-

Sandstone. (© Richard Pieper)

mented together with a variety of minerals (silicates, iron oxides, limonite, calcite and clays). These cementing materials make one sandstone behave very differently from another. Sandstones containing silica are quite hard, strong and decay resistant, whereas those containing calcite resemble limestone in their susceptibility to acid damage, and those containing clay absorb water and deteriorate more easily. Brownstone is the common name for brown, red and purple sandstone, particularly the arkosic sandstone found in the Triassic basins of the eastern United States and widely used in the Northeast in the second half of the 19th century. Because they have a granular texture throughout, sandstone surfaces stay matte even when worked.

Most sandstones tend to form a hard outer crust that retains dirt; thus, a simple, although not foolproof, way to differentiate them from limestones is to check the condition of a building's ledges exposed to the rain. They would generally be the dirtiest part of a sandstone building but the cleanest part of a "self-cleaning" limestone building.

Most sandstones have a tendency to absorb moisture and do not withstand frost action well, so they should not be used as foundation stones and should be protected from excessive moisture.

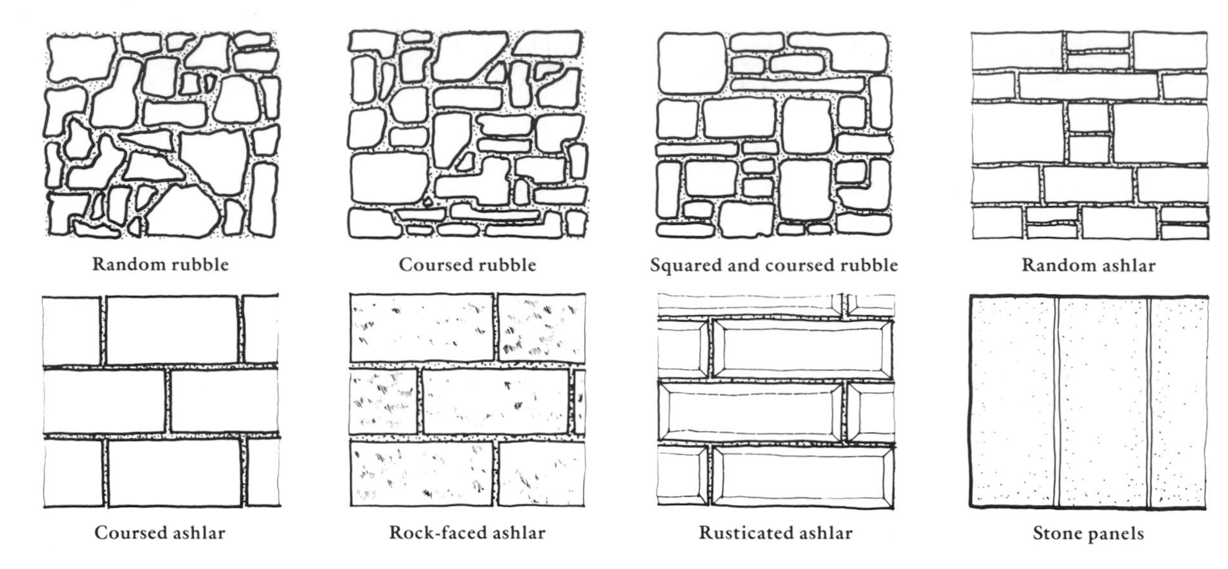

Common stone bonding patterns. (Robert D. Loversidge, Jr., AIA, from the *Old-Building Owner's Manual,* Ohio Historical Society, by Judith L. Kitchen, 1983)

Granite

Granites are dense and hard, resist wear and usually have a speckled appearance. Although geologically the term "granite" is restricted to certain crystalline rocks of igneous origin made up of quartz, feldspar and mica, in the stone and construction industry the term is used for almost all igneous rocks with an interlocking granular texture; for example, "black granite" refers to black, fine-grained igneous rocks such as basalt and diabase. Because they have low porosity and permeability, granites have a high resistance to weathering. Granite was often traditionally used for features in contact with the ground (foundations and steps) in predominantly limestone and sandstone buildings. While most granites are quite uniform and durable, some are susceptible to weathering and iron staining.

Marble

In the stone and building industry, the term "marble" refers to any rock that can take a high polish, including a few limestones and granites. Geologically, "marble" refers to certain crystalline rocks composed mainly of calcite or dolomite — metamorphosed limestone. The purest form is white statuary marble; in others, impurities produce a riot of colors and patterns. Because it is not as hard as most other construction stones, marble can be worked more easily and is used mainly for ornamentation. Marbles are often very sensitive to damage by atmospheric pollutants, and many are not suitable for exterior use. This is why the marble monuments of ancient Greece and Rome are rapidly deteriorating in today's polluted urban environment.

Brick

Beginning with early civilizations, the principal raw material used in traditional brick making was clay found in surface deposits. Surface clay was gradually replaced in the 19th century as techniques were developed to crush and pulverize shale, making it into a clay as well.

Clays are complex minerals composed of silica and alumina, as well as iron oxide, calcium oxide, lime, magnesia and other substances. From one locality to another the proportions of these minerals varied and contributed to color differences, for example, making the bricks of one city more purple than those of another.

To reduce shrinkage, warping and cracking as the brick dried, sand was usually added up to a sixth or even a third of the total volume, generally reducing the brick's strength. Because the basic raw materials for making brick — clay, sand and water — were available almost everywhere, brick was made locally, sometimes even at the building site itself.

Handmolding brick. (Jack Hunt, Old Carolina Brick Company)

Making Brick

Manufacturing bricks, whether in Babylon 6,000 years ago or in Illinois today, has always consisted of five tasks: removing the clay from the ground (or removing the shale and pulverizing it), mixing it with sand and water, shaping the mixture, drying it and, finally, firing it.

Surface clay was dug out of fields with claylike soil, first by hand and more recently using graders and mechanical shovels. Rocky clay or shale was also removed from escarpments and outcrops with explosives or mechanical equipment. Shale was left to weather for several months to a year so that it would partially disintegrate and be easier to crush.

After rough extraction, clay, sand and water were mixed in a process called tempering. In earliest construction this was done simply with a spade in a soak-pit in the ground; later, mixing was accomplished mechanically.

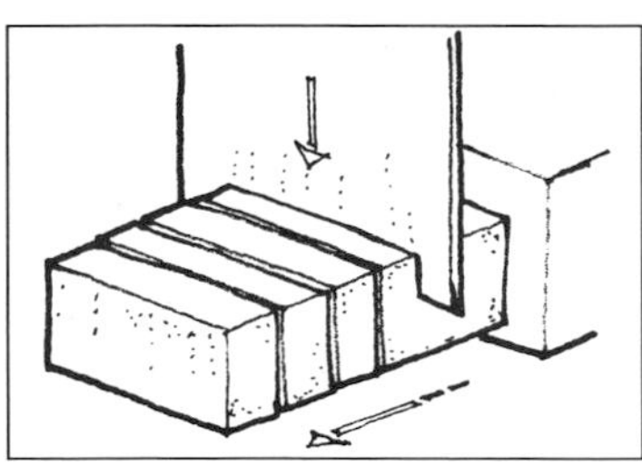

Top: Clay-cutting machine of the Interstate Brick Company, Salt Lake City. (Jack E. Boucher, HAER)

Above: Extrusion method. A continuous bar of clay is cut into brick-sized slabs. Slabs are textured before firing.

Shaping

Three methods have been used to transform tempered clay into a brick shape: molding, pressing and extrusion.

Molding, the soft-mud method of shaping bricks, involved having a laborer knead soft clay by hand, tamp it into a wooden mold that had been sprinkled with sand or water to prevent the wet clay from sticking, leveling the top surface with a wooden or steel strike or straight edge, and then withdrawing the brick from the mold. The result was a low-density irregular brick that might well bear the workman's fingerprints. If sand had been used on the mold, the texture of the brick was satin (with fine sand) or sandy; if water was used, the surface was smooth. Because the clay had not been compressed, the brick was not very dense. Most buildings built before 1875 probably used handmolded brick, which continued to be made well into the 20th century.

Pressing, the dry-press method of brick shaping, mechanized the process of filling the molds. Clay, which did not

Royal River Brickyard, North Yarmouth, Maine, no longer in business and one of the last scove kiln operations in the United States. Manufacturing brick by this method was most popular in the 19th century.

Excavating glacial deposits of sand and clay suitable for brick. Old brickyards often wcrc locatcd ncar dcposits until the clay was exhausted. (Philip Marshall)

Tempering the clay by mixing clay and water in a power-driven brickmaking machine. This equipment, largely unchanged in the past 100 years, was the only specialized capital purchase for small operations such as this. (Philip Marshall)

Molding the bricks by pressing soft, tempered clay into wood molds. The bricks were "water struck" as each mold and palette was soaked in a barrel of water (left) before being positioned under the press, where clay is introduced. In the distance the excess brick is removed with a "strike" (board). The bricks then are delivered to a motorized conveyor. (Philip Marshall)

Transporting raw bricks to the "hacks," roofed shelves that protect the bricks from rain while allowing them to dry slowly. (Philip Marshall)

Moving unfired "green" bricks after several weeks of drying to the kiln site, where an intermittent scove kiln is being constructed by using the green bricks it will fire. It was "intermittent" because the kiln was heated up and cooled down each time new bricks were fired. (Philip Marshall)

Scove kiln. When complete, this kiln contained more than 30,000 green bricks, arranged to permit fuel (wood, gas) to enter through arched spaces at the base and allow heated gases to circulate throughout. The kiln was carefully fired for several days, cooled and completely disassembled. Its contents were graded (sorted) by the size, color and quality of each brick. (Philip Marshall)

have to be as wet as with the soft-mud method, was pressed into the molds with plungers that were first handoperated and later steam powered. Some molds were designed to make a "frog" (depression) in one face of the brick to improve its bonding with the mortar and also give manufacturers a place to inscribe their names. Pressed bricks are more regular than those made by hand; also, mechanical pressing compressed the outer surface, increasing its ability to resist weathering. Machine-driven presses became common in the last quarter of the 19th century; pressed brick is still made today.

Extrusion, the stiff-mud method of brick shaping, facilitated continuous production of bricks and is now the most common brick-making method. It involves extruding a stiff clay or powdered shale mixture through a die under pressure. The resulting bar of compressed material is then sliced by steel blades or wires into individual bricks. This method allows adding a texture to the surface by passing the bricks under combs or pressing them with patterned cylinders. It also enables production of bricks with holes or in special shapes such as curves for building corners and columns. In the early 20th century extrusion became the chief method of modern brick manufacture.

Early kiln in Williamsburg, Va. (© Michael Devonshire)

Drying

After shaping, bricks must be air dried; otherwise, in a kiln they could warp or explode because of a buildup of steam. In early brick making, drying was done in a few days in the sun. Later, shelters or drying sheds were used. Fires in sheds speeded drying in humid or wet climates.

Firing

The constituent materials of brick are partially sintered (fused) when a brick is fired (burned) in a kiln to produce a hard material resistant to the effects of water. The quality of the firing depends on the evenness of the temperature, which should average about 2,000°F, and the duration of firing. Firing is critical in determining a brick's strength, hardness, color and porosity.

The earliest firing operations used piles of bricks called clamps, built on the construction site from unfired bricks and covered with clay or earth. The cavity was filled with more bricks stacked in alternate layers with wood that was then set

Brick bonding patterns, top to bottom: English, Flemish, American (common) and stretcher (running) bonds. (Paul Kennedy)

on fire. A burning took several days, and the resulting bricks were quite uneven because of variations in temperature and the effects of smoke. A single firing could produce a full range of types, from underfired salmon-colored bricks to overburned ones fused into a vitrified mass or clinker. Between 10 and 40 percent of the bricks might be totally useless. These handmade temporary kilns were used until the end of the last century.

Permanent kilns were built of fire brick, usually near plentiful sources of clay or shale where demand was high. The firing process was similar to that of the on-site kilns, although a number of technical innovations improved efficiency, consistency and quality.

Tunnel-shaped kilns that fired continuously came into use early in the 20th century and virtually replaced all other types. The new kilns could be fed constantly, and they produced much larger quantities of high-quality, uniform bricks.

Characteristics of Brick

In general, the more a brick is fired, the stronger, less porous and often darker it becomes; color is thus a good indication, although not a certain one, of the quality of traditionally made brick. Several major categories, from hardest to softest, can be distinguished. Although earlier kilns might produce all types in one firing, with later kilns special firings were made to produce each type. Some of the most common types of brick are:

- Fire bricks from near the center of early kilns, often dark brown or black and used to line chimneys, flues and kilns
- Paving bricks, usually brown and used for exterior paving
- Face bricks, regular, well-fired, generally red bricks that are hard, uniform, weather resistant and attractive and used on building exteriors
- Common bricks, poorly fired, often pink or orange and used for interior walls or as nonstructural fill

As mentioned, brick color is not always a guarantee of the brick's strength. Some underfired bricks may appear dark — and, therefore, strong — because they have black interiors from impurities in the clay, and some fire bricks — one of the strongest types — are white or pale yellow, not dark.

Face bricks in a highly decorative pattern on a facade. (Dinu Bumbaru)

Strength

Like stone, brick is strong mainly in compression. A brick's strength lies in its better-fired outer crust. A long-used test is to score the surface with a steel knife blade; if it scratches easily, it may not be hard enough to use as a face brick. Current standards call for 2,500 psi (pounds per square inch) for standard face brick and 3,000 psi for face brick used in a highly exposed area.

Water absorption

The more the clay is compacted, and the better the brick is fired, the less porous the brick is. The fused outer skin of the brick is its main line of defense against water. Water can penetrate to the more porous interior when this hard outer surface is removed by sandblasting. Water can also penetrate through cracks in the brick that may result from structural deformation in the building — for example, settlement — or because mortar joints are too rigid to absorb vibrations and shifts in the building. In cold climates, water within the brick can freeze, forcing the surface of the brick to spall, further exposing the soft interior to rapid deterioration.

Dimensions

Bricks are a versatile building material. Their small size prevents them from cracking and warping when drying and firing, allows great flexibility in design and permits a single bricklayer to work alone. Yet when bonded together with mortar, they create a strong wall.

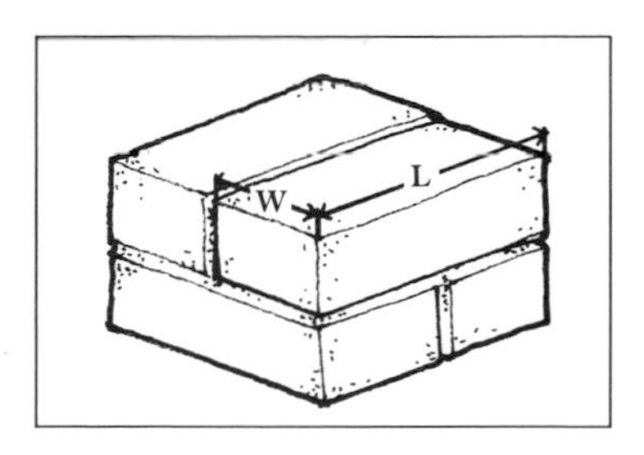

Brick modular unit.

Bricks are modular units. In principle, the length of a brick is equal to twice the width plus a joint, or triple the height plus two joints. Until 1899 brick sizes varied widely from one area and manufacturer to another. Then the U.S. National Brickmakers Association standardized the size of face bricks to 8⅜ by 4 by 2⅜ inches. Common bricks were slightly smaller. Some variation still exists, however. A double thickness wall (referred to as having two wythes) is a little more than 8 inches thick, and one that has three wythes is almost 13 inches thick.

Diagnosing Masonry Problems

In dealing with masonry problems, applying Band-Aid solutions to the most visible evidence of a problem is not a sure cure. It is necessary first to identify the problems, their nature and extent; then to assess the underlying cause of each problem; and only afterward to decide on a treatment, which will arrest the process of deterioration, not merely treat the symptoms. Inaccurate diagnosis can lead to the selection of ineffectual or even harmful treatments. The diagnosis should be carried out by an unbiased investigator whose interest is the protection of the building, not someone trying to sell a particular product or service.

The diagnosis of masonry problems likely begins for one of two reasons: the building owner has called in someone to treat a specific problem, or an architect is undertaking a general rehabilitation of the building.

An expert may be able quickly and easily to identify straightforward problems, such as leaky gutters or a cracked window sill in a house or small building. But more severe or complex problems may require more detailed investigation and data collection, often including measurements of the moisture content of the materials, laboratory analysis of mortar or masonry units and monitoring of temperature and humidity over a period of time.

If the building is large, or if the problems are complex, as much time as a year may be needed to complete an accurate diagnosis, based on conditions in different seasons and to avoid rushing into useless or harmful treatments. This is particularly important for significant buildings in which protection of the original fabric is of paramount importance.

Diagnosing problems at Strathmore Hall (1902, 1914), Rockville, Md. An architect checks soundness of an earlier repair (opposite) and looks for damage at the roofline, demonstrating how water from the roof has damaged masonry and wood (right). (Paul Kennedy)

Diagnosis at a Glance

Problem	Possible Solutions
Bulges	
In walls	Anchor wall to structure. Rebuild wall.
Cracks	
Due to unstable soil	Improve drainage. Cut trees. Extend foundations.
Due to unstable foundations	Repair or rebuild foundations.
Due to unstable structure	Repair or replace structure or walls.
In brick	Replace.
In stone	Fill with grout, repair with epoxy or replace.
In mortar joints	Repoint.
Deteriorated or Broken Masonry	
In general	Find and solve cause.
Minor, superficial damage	Consider leaving as is.
Bricks	Replace.
Stones	Reattach with epoxy. Apply composite patch or undertake mechanical repair.

Weakness in an arch. The repointed cracks indicate the arch was not strong enough to hold the weight above it. (Paul Kennedy)

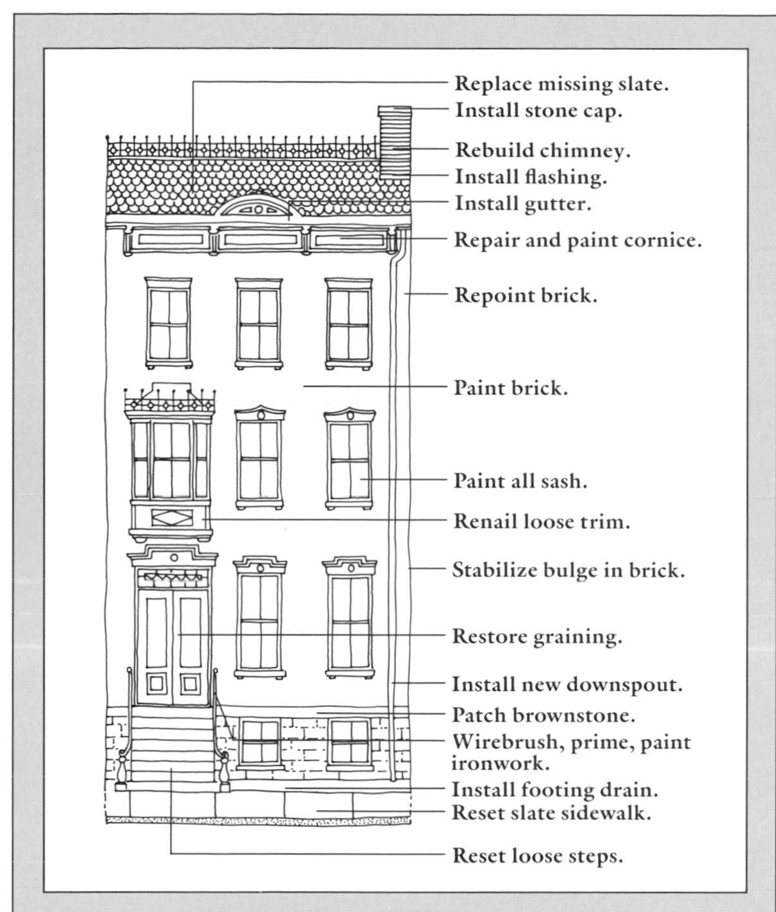

Repair at a glance. A structure may require any or all of the indicated work on its masonry and nonmasonry features to return it to a good state of repair. (Preservation League of New York State)

Deteriorated or Missing Mortar

In general	Repoint joints between brick or stone units.

Dirt or Stains

If not threatening structural integrity	Consider leaving as is.
If threatening structural integrity or unsightly	Choose cleaning method based on type of dirt and masonry.

Moisture Penetration

Due to roof, window leaks	Repair roof or window.
Due to rainwater (rare through masonry units)	Apply paint or water-repellent coating.
Through foundations	Improve drainage. Waterproof foundations.

Why Does Masonry Deteriorate?

Brick and stone are subject to attack by a host of forces. How successfully a building resists these pressures depends on how well it is designed. A well-built structure may withstand these forces indefinitely. But buildings with weak foundations and parts that do not shed water or absorb movement and those made of inferior brick or stone will deteriorate rapidly.

The leading causes of decay are

- Excessive moisture within the masonry that gives rise, for example, to destructive crystallization action of soluble salts as well as freeze and thaw action in northern climates
- Water flowing through walls that can lead to differential settlement, deterioration of adjacent materials (rusting iron anchors or rotting wood, for instance) and other structural problems
- Inappropriate rehabilitation techniques such as sandblasting

Secondary factors are:

- Abrasion by the wind and windborne solids
- Differential expansion that places internal stresses on the building when one part responds to thermal stresses more than another or when a building shifts because of weaknesses in the soil, foundations or structure
- Mechanical impact caused by accidents (for example, a falling tree), wear and tear by users or some renovation techniques
- Chemical disintegration caused by pollutants in the atmosphere

Two causes of critical masonry deterioration: freeze-thaw cycles (top, NTHP) and excess moisture from floods (above, U.S. Coast Guard).

Right: Water-and-soap cleaned surface compared with sandblasting. (Dinu Bumbaru)

Far right: Effects of thermal expansion. Over time, one wall has warmed faster than the other, causing the crack. A caulking material has compounded the problem. (Ward Bucher)

Acid rain, which causes chemical disintegration, is particularly severe in the northeastern part of the North American continent where prevailing winds bring pollution from inland coal burning and industrial operations. The sulphuric acid in acid rain turns limestone into gypsum, which washes away. Chemical disintegration also may be caused by organic matter such as moss and lichen, which secrete acids.

Top: Mechanical impact, a secondary factor in masonry deterioration. (National Oceanic and Atmospheric Administration)

Above: Destruction caused by salts. In this unusual example, salt built up in the stone columns, buried for hundreds of years, began to cause the material almost to melt away after the stone was uncovered. (© Richard Pieper)

Key Steps in a Diagnosis

Before undertaking any masonry work, it is essential to evaluate how significant the building is architecturally and historically and particularly how important the masonry is to the building's overall character. This permits selection of repairs that maintain the integrity of the building and do not damage the building elements and materials that are important in defining its character. The following steps allow a thorough diagnosis of masonry problems. Time constraints or other factors, however, may require changing the order or combining some of the activities.

Historical Research

Even if the building in question is a home without outstanding architectural or historical importance, an owner or architect first can gain a useful understanding of its design and construction by spending a short time researching its history.

Begin with the previous owner, who already may have much information on the structure's history or can at least suggest where to find it. Neighbors and other people knowledgeable about the history of the community may be helpful, although memories are not always completely reliable. The building permit department of the municipality or county may have the house plans on file. Old tax rolls, deeds, account books, historic maps and city directories may also give useful information on when and by whom the house was built. For important public buildings, information may be found at the local library, public archives, and university libraries, history, art and architecture departments.

Research may help identify the type of stone or brick used

A Minneapolis restoration. Efforts to restore this 19th-century residence (right, Bruce Goldstein) led to historic photographs such as the one above (Elizabeth Schutt), which indicated the house originally possessed a limestone masonry porch.

The porch (right) after restoration with matching stone in 1985 and family and friends (above) on newly restored porch. (Bruce Goldstein)

and may even reveal the type of mortar and construction details. It could also identify other buildings built during the same period by the same architect or builder, who may have used similar designs and materials. This, particularly, may lead to a useful line of investigation. Have similar buildings developed the same problems? How were they dealt with and with what success?

Also, try to find out how the building has been used, how it has been repaired or changed and whether it was ever subjected to unusual conditions. These might include being left empty or unheated, suffering a fire or housing a function with high interior humidity (such as a museum or laundry).

A fair amount of historical research on a simple building can be carried out by a dedicated amateur, but a competent and knowledgeable researcher should do a thorough investigation before the rehabilitation of a historically significant building. This investigation should include not only a historical and stylistic analysis but technical research as well.

Local Regulations

Find out what local zoning and building regulations and fire codes affect the building. These regulations may have changed since construction, and the building may not meet present requirements. A nonconforming condition may be "grandfathered" (allowed to continue), or the municipality may call for the building to be brought up to current standards if it is to undergo extensive rehabilitation.

Check whether the building has been designated as historic or is in a historic district and what special regulations or guidelines apply. If an application is being made for grants or tax credits, find out what the criteria are. They may well be stricter or different from municipal regulations.

Plans and Photographs

A good set of photographs of the affected area can be useful in locating and identifying problems. While subtle color differences visible on color film might help clarify problems, fine-grain black-and-white film offers the best image definition. In either case, the pictures should be taken with the camera as close to horizontal as possible, from a window in a building across the street if necessary, to prevent distortion. If possible, photos should be enlarged to a recognized scale (for example, ¼ inch to 1 foot) so that lengths and areas can be measured, or at least estimated, directly from the photos. In addition to basic shots, a telephoto lens can be used to record details of problem areas. All these photos can then be copied, to be used first for notes during the inspection, then to outline work to be done. In fact, photos indicating which areas are to be repointed or which bricks or lintels replaced can form part of the legal contract with the building renovation contractor.

If an extensive rehabilitation including more than masonry work is being planned, a good set of measured drawings of the building is invaluable. Drawings make it easier to analyze the construction and use of the building and often suggest opportunities that are not evident from looking at the building directly. If it is not possible to find plans, an architect or architecture student can be hired to make them. The plans should be drawn in large enough scale to work on — at least ⅛ inch to 1 foot, and preferably ¼ inch to 1 foot.

It is best to have photos and drawings before actually doing the detailed inspection.

Carlyle House (1752), Alexandria, Va. Constructed of a very soft local sandstone, this historic property required restoration in 1976. (Richard Bierce)

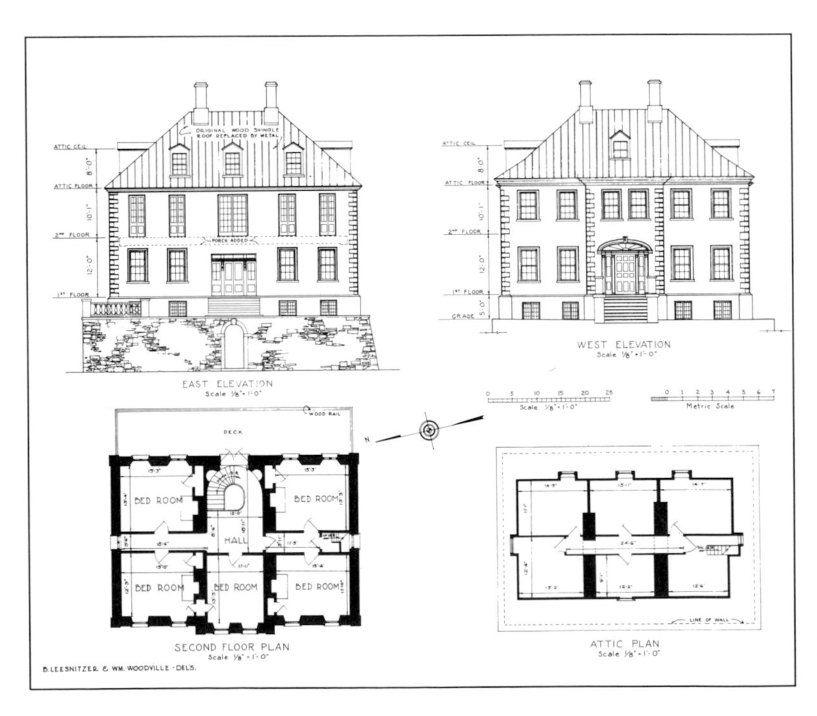

Carlyle House (above, NTHP) before restoration and measured drawings (right, HABS) that helped guide the restoration project.

Basic Inspection

The following is a description of how an architect, preservation consultant or building inspector would carry out a complete inspection of a building before undertaking an extensive repair or rehabilitation program. Inspection of a simple house without particular problems might not involve every aspect but would follow the same basic approach outlined here.

The aim of an inspection is to discover all existing and potential problems as well as to identify possible causes and potential solutions. The problems can range from major structural defects such as out-of-plumb walls, bulges and cracks to minor problems such as missing, damaged or stained stones, bricks or mortar joints.

Scaffolding, a useful but expensive way to conduct an inspection. The costs may be written off by using the scaffolding for other rehabilitation work. (NTHP)

Access

For a complete inspection, the inspector should be able to get close to every part of the exterior wall and should be able to touch it while relaxed; it is hard to be attentive to masonry problems while hanging over the edge of the roof on a rope or dangling from a crane. A ladder should work for small buildings; for larger ones, a cherry-picker (rigid-arm crane) or a swing stage used for cleaning windows is useful. The services and equipment of licensed riggers can often be rented by the day. Scaffolding is the most solid, and the most expensive, base from which to inspect, but it may be hard to justify this expenditure before construction begins.

Safety measures

Here are some key safety measures that anyone inspecting a building should observe. These also apply to the actual rehabilitation work. Although the precautions seem obvious, they are too often ignored.

Do not take chances. If uneasy at heights, get someone else to do the inspection. The most serious building problems are often in the most hard-to-reach, and therefore neglected, areas such as roof flashing and parapets.

Dress appropriately. Wear antislip shoes, preferably safety shoes or boots with steel toes. Tightly fitted clothing and a hard hat are also essential. Goggles are vital for any in-

The Inspector's Tool Kit

In addition to a checklist, base plans and photos on a clipboard, lots of notepaper and a good supply of pencils and pens, the inspector should have the following items:

- Tape measures, ideally a 25- or 50-foot tape for overall dimensions and a 10-foot tape for details
- Camera and flash, to take close-ups of particular problems, and a telephoto lens for details
- Flashlight
- Plumb bob, a lead or brass weight at the end of a string, to determine whether a vertical part of the building is a true vertical
- Carpenter's level, to determine whether the horizontal parts of the building are true
- Binoculars, to examine the upper parts of the facade without climbing
- Handheld microscope, monocular or magnifying lens (10 power)
- Rubber mallet
- Ice pick or thin-bladed instrument to probe wood
- Small crowbar
- Thermometer and humidity gauges
- Moisture meter

spection or construction work that could possibly involve chipping or shattering brick or stone.

Secure ladders and scaffolding. Ladders should rest solidly on the ground and against the wall. The distance between the foot of the ladder and the wall should be about one-quarter of its length. Do not carry equipment up the ladder or scaffolding; pull it up on a rope after you are safely up and well secured.

Secure scaffolding to a solid part of the building. Make sure that the scaffolding does not damage the building. The ladder should be enclosed within the framework of the scaffolding and should be staggered from one level to the next. When working high off the ground, use special safety equipment such as ropes and belts. Perhaps the greatest danger is power lines; the consequences of touching one, when moving a ladder or crane, can be fatal.

Use the safest inspection method. Inspect window and door openings from inside if possible. If it is necessary to lean out of a window or over the edge of a flat roof, the inspector should be well secured. An amateur should never attempt these tasks.

Steps to follow

Inspectors should take their time and proceed methodically. A careful inspection of a house will normally take half a day; a major building could take weeks. Building owners may wish to carry out their own preliminary inspection to make themselves familiar with potential problem areas when they later accompany the professional inspector on the full inspection.

Usually, inspectors start outside, checking the general site conditions. Then they examine the exterior walls, starting at one corner and working their way around the building. They will look first at the general condition of the walls, noting structural problems, bulges or major cracks. Then they study the masonry units, assessing the condition, problems and possible remedial action for each stone or brick, at the same time examining the masonry joints for materials and conditions of pointing. Indoors, inspectors will conduct a detailed room-by-room review including basement, crawl space and attic. Finally, they check out the roof.

A key part of the inspection is figuring out where rainwater goes. Water is a building's main enemy, and every part should be designed to shed it. The best time to inspect is while it is raining; check for leaks and water accumulation on different parts of the building and on the ground.

The inspector should write down everything carefully, noting locations of problems on close-up photos or plans. Properly recorded data will always repay the extra time involved. The date and weather (including temperature and relative humidity) at the time of the inspection should also be noted.

On enlarged copies of the photos or drawings, the different materials and where the masonry and mortar joints have deteriorated should be indicated. Using a code, map where

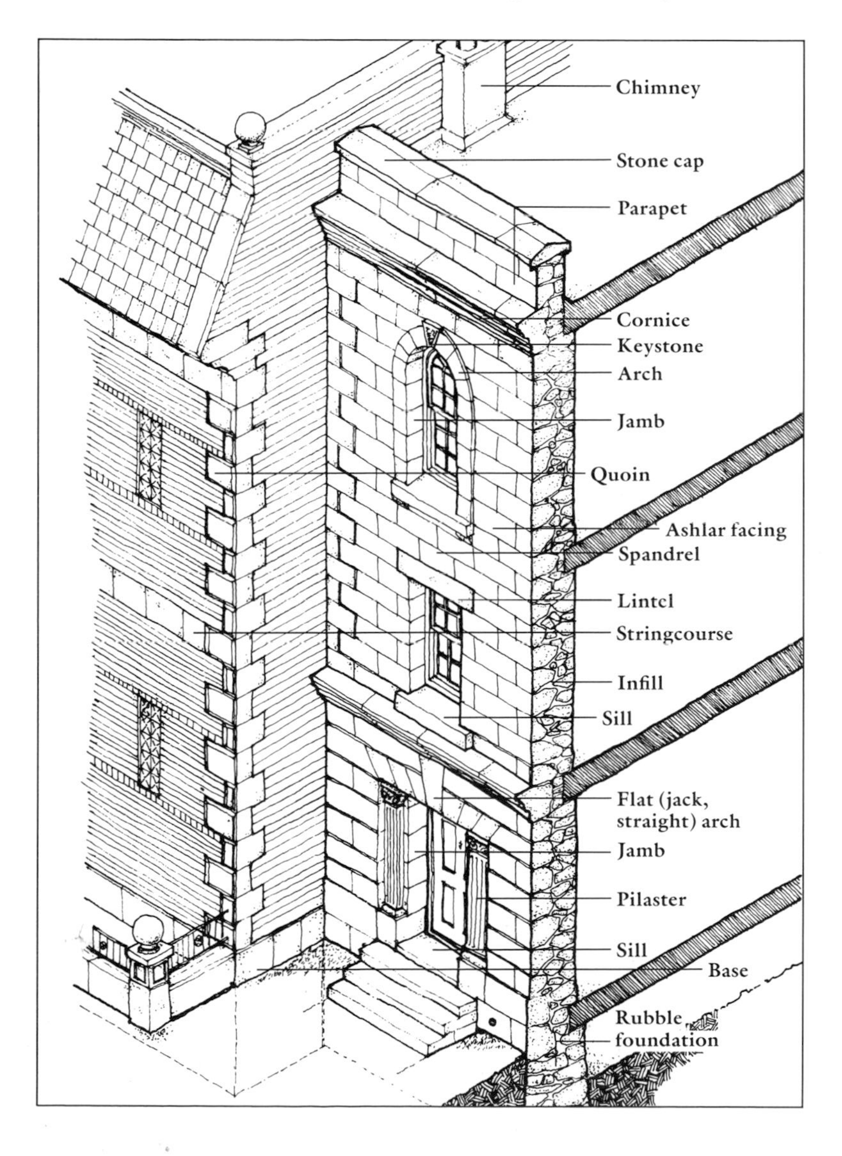

Parts of a masonry building. Familiarity with basic architectural nomenclature is always useful in a diagnosis.

Townhouses in need of attention. In an inspection, all details of the exterior should be noted, including condition of stairs, railings, paint and ornamental work. (P.C. Turner Associates and Nicholas Satterlee and Associates)

the masonry units are significantly deteriorated (that is, where their failure seems to threaten the structural integrity of the building); where they are moderately deteriorated (for example, where there is no actual threat but visible damage that should be monitored for future action); and where the bricks or stones are sound. The same procedure should be followed for mortar joints. Areas where repointing is needed should be indicated on the drawings or photos. Also mark major cracks and areas of bulging. Note areas that are heavily soiled or that have unusual stains. And note features that have delicate detailing that should be protected during cleaning.

Every surface, outside and in, should be looked at carefully — ceilings, walls and floors — giving particular attention to joints between various elements such as door and window frames and sills, as well as roof flashing. Building owners should use the checklist at the end of this chapter to help them detect various masonry problems and symptoms.

Diagnosis

At this point, an experienced investigator should be able to make an accurate diagnosis of most straightforward problems and be in a position to select remedial treatments. For example, interior dampness may occur in the vicinity of clogged downspouts or earth piled against the building. In such a case, the apparent source of the problem should be corrected and the situation monitored to make sure that the problem clears up.

But where the nature and causes of problems are not readily apparent, as is often the case with moisture problems, further investigation is necessary before a diagnosis can be made. For example, a "tidemark" line of salt deposits near the base of the building may indicate the presence of excessive moisture in the wall. But the source of the problem (rising damp, rain back-splash, condensation) cannot be identified and treated without expert detailed examination (see Finding and Treating Moisture Problems).

Detailed Examination

In cases for which a basic inspection is unable to identify the cause or causes of complex problems, selected physical examination, tests or long-term monitoring may be required. At this point, an inspector or architect may call in a specialized restoration consultant.

Specific tests

The following are some specialized tests that can be carried out to ascertain causes of masonry deterioration.

Tests for salts determine the nature of salt deposits on the surface of, or within, the masonry units and may help identify the sources. For example, phosphates and nitrates might come from fertilizers, chlorides from de-icing salts.

Tests for the composition of materials determine the physical and chemical nature of the masonry units and the mortar.

Soil borings determine whether ground water presents a problem. For example, in houses and other small buildings, small test pits are dug during the year near the foundations to the depth of the footings to determine the ground-water level.

Moisture meters monitor and assess the moisture level in materials.

Hygro-thermographs monitor temperature and hu-

Rear of the Morse-Libby Mansion, Portland, Maine. Moisture damage is evident from peeling mastic, a mid-19th-century coating made of sand and oil or varnish. (Richard Cheek)

midity levels over a period of time and preferably through a winter and summer.

Movement of cracks cannot be detected with ordinary measurements so a variety of tests has been devised: using special instruments that indicate the magnitude and direction of crack movement; plastering a small part of the crack to see whether it breaks; fixing a tube of glass across the crack to see if it breaks; or attaching a brass strip on each side of the crack, scoring a fine line across the strips and measuring its movement over time.

Physical probings of the building verify construction details and their state of deterioration. If samples of masonry or mortar are to be taken or physical probes made, these should, if possible, be done on hidden parts of the building and on already badly damaged parts that would have to be replaced anyway.

Infrared thermography measures heat transmission through a wall, indicating areas of air leakage as well as moisture and heat loss; sometimes thermography can give an X-ray-like indication of the structure of the wall (for example, showing the location of columns behind wood siding).

Efflorescence can be analyzed by brushing it off and watching whether it returns in two weeks. This may indicate how active the problem is, although a wall can continue to effloresce for quite a while after the basic moisture problem is solved or may fail to effloresce when still damp and salt laden (see Finding and Treating Moisture Problems).

Repeating inspections

For problems that remain unresolved, it may be necessary to repeat the basic inspection and detailed examination in six months and perhaps again a year later. This will give a better overall picture, especially with moisture problems that may change during periods of high humidity, heavy seasonal rainfall and the condensation and freezing of water vapor coming from the building's interior during wintertime in northern climates. Further examination also permits monitoring of the effects of treatments undertaken after the earlier inspection. Even after all the problems and causes have been identified and solved, periodic inspection should be part of the regular maintenance program of every building. This should take place every five years — more often for parts particularly exposed to the weather.

Inspecting for Masonry Problems

The Site

Look for	Possible Problems
Environment	
General climatic conditions, including average temperatures, wind speeds and directions, humidity levels and average snow accumulation	Severe conditions can lead to masonry deterioration, including cracking, spalling and efflorescence.
Number of freeze-thaw cycles	Severe cycles can produce damage from frost action.
Location near sea	Salt in the air can lead to efflorescence.
Acid rain in the region or from nearby industry	Acid rain is particularly harmful to limestone, marble and sandstone.

Puddling. Unless corrected, this condition can lead to deterioration of the mortar and the masonry itself. (Paul Kennedy)

Look for	Possible Problems
Proximity to a major road, highway or railroad	Vibrations are harmful to mortar joints and other building parts.
Proximity to a major excavation	Deep excavation (for foundations of a large building, deep sewage system, subway) can lower water tables and alter drainage, thus drying up soil (especially clay), possibly leading to foundation failure.
Location in the flood plain of a river, lake or sea	Floodwaters can bring damaging moisture to foundations and walls.
Exposed or sheltered sections of a building	Exposure to the sun and elements affects moisture evaporation and rain penetration.
Terrain	
Soil type (clay, sand, rock)	The type of soil influences water drainage around the structure. Excessive water in the soil can cause rising damp, leading to structural problems.
Slope away from the building on all sides	If no slope exists, puddles will form at the base of the walls during heavy rains, leading to water penetration.
Earth covering part of a brick or stone wall or planter boxes against walls	Moisture accumulation or penetration is possible.
Asphalt or other impervious paving touching walls	Water accumulation and rain back-splash onto the walls can result.

Look for	Possible Problems
Trees and Vegetation	
Species of trees within 50 feet	Elms and some poplars dry up clay soil, leading to foundation failure.
Branches rubbing against a wall	Branches abrade surfaces and obstruct gutters.
Ivy or creepers on walls	Leaves prevent proper drying of the masonry surface. Tendrils from some species can penetrate mortar joints.

The Building

Look for	Possible Problems
Overall Condition	
General state of maintenance and repair	A well-maintained building should require fewer major repairs.
Evidence of a previous fire or flooding	Such damage may have weakened structural members or caused excessive moisture.
Signs of settlement	Uneven settlement can crack foundations or walls or result only in sloped or wavy mortar joints.
Vertical walls	Crooked walls may be a sign of stabilized building settlement as well as unstable foundations and possible building collapse.
Roof	
General condition	A leaking roof can leave stains on upper-story walls and ceilings as well as damage masonry.

Look for	Possible Problems
Copings on parapets; flashing on cornices	If copings do not extend well beyond the wall, they cannot protect the upper wall from rain damage.
Gutters, downspouts (front and back) and roof drains	If missing or in poor repair, not directing water away or clogged with vegetation, these permit water to accumulate and penetrate.
Flashing and metal capping, especially on flat roofs	Rusting metal can allow water to penetrate.
Windows and Doors	
Straight openings	If deformed, this is a sign of building settlement.
Sills sloped to shed water; drips under sills to prevent water from running back underneath; caulking	If any of these is inadequate, water can penetrate into the wall.
Basement	
Composition of foundation walls	Stone or brick is more likely than concrete to allow water to infiltrate.
Water condensation or other signs of moisture	Wood joists resting on foundation walls may begin to rot at the ends. Termites or algae may be present, causing damage to wood or masonry.
Damp-proof course	This can impede rising damp, lessening deterioration of the masonry wall.
Interior	
Bathrooms, kitchens, laundries, humidifiers	These are sources of excessive and unwanted moisture, leading to condensation and deterioration.

Look for	Possible Problems
Cracked plaster, signs of patching, floors askew	These are signs of building settlement.
Damp walls, stains on walls, rotting wood	These indicate water infiltration.

Building ghost. The exposed bricks are most likely a soft brick rather than a hard-fired exterior face brick. (© Michael Devonshire)

Look for	Possible Problems
Walls	
Construction method (solid or cavity, load bearing or not load bearing)	Knowing how a wall is constructed will help in analyzing problems and selecting appropriate treatments.
Evidence that parts of the building were built at different times or of different materials	Similar problems with various parts may need different treatments because of different materials.
Weep holes (small holes at the bottom and top of walls)	Holes allow ventilation from the air space in a cavity wall. If missing, they can be added during rehabilitation.

The Masonry

Look for	Possible Problems
Materials	
Composition, including secondary materials; characteristics (color and color variation); texture (smooth or patterned surfaces)	Types of materials indicate susceptibility to damage and should be matched if brick or stone is replaced.
Areas of delicate carving or fine moldings	These sections may need special attention or protection during rehabilitation.
Missing or broken bricks or stones	Missing material may allow water penetration.
Evidence of sandblasting, such as a pitted surface; evidence of erosion, crumbling, flaking, scaling or spalling	Surface deterioration can be aesthetically displeasing and damaged surfaces allow water to penetrate.
Dirt or stains	Surface stains usually cause few problems other than being unpleasant to look at.

Look for	Possible Problems
Bulges and Cracks	
Bulges	Bulges indicate that the wall has moved and corrective action may be necessary.
Outer-face bulge	Solid walls tolerate movement less if only the outer face is moving; immediate remedial action may be necessary.
Cracks	Cracks indicate movement within the wall. Small cracks may be patched; large cracks may require reconstruction of the affected area.
Enlarging cracks	Active cracks indicate a continuing problem. The cause must be dealt with before the crack itself is repaired.
Moisture	
Water penetration through joints between masonry and other building components, through masonry joints or, rarely, through brick or stone units	Moisture can lead to deterioration of the masonry and other parts of the structure.
Staining or white deposits (efflorescence) on exterior walls	White deposits are evidence of excessive dampness. Efflorescence on most new or newly repointed walls (new construction "bloom"), however, is natural and will disappear after normal weathering.
Location and type of salt deposits on surface	Deposits can indicate a source of dampness, such as rainwater or ground water, inside the building materials.

Keystone in a brick arch. Keystones originally played a major structural role in arches; most are now primarily decorative. (Tony Wrenn)

Look for	Possible Problems
Coatings	
Paint; type of paint	A paint that does not breathe can trap moisture within the masonry and cause the surface to spall.
Blistering, flaking and peeling paint	These conditions indicate the paint does not breathe.
Waterproof or water-repellent coating	Such coatings trap moisture within the masonry.
Mortar Joints	
Type of mortar (lime based, usually whitish; or portland cement, grayish and very hard)	A cement mortar is too hard for old masonry and can lead to cracking or other damage of the brick or stone units.
Condition of mortar (crumbling, eroded, missing)	Damaged or missing mortar can allow moisture to penetrate; repointing may be required.
Broken or chipped edges of brick or stone along joints	Damage may indicate mortar in joint is too hard.

Maintenance Checklist

Good maintenance is the best way to care for a building. A well-built and properly maintained building may never need extensive rehabilitation. At the very least, maintenance can slow deterioration and postpone repair or replacements.

Routine maintenance of a building should include inspecting and dealing with minor problems on a regular basis — annually for large or important buildings and every few years for smaller buildings in good condition. The following steps are recommended.

Survey the landscaping. Make sure that the site is landscaped so that water drains away from the building. Ensure that there are no trees whose roots are too close to the foundations or whose branches can rub against the building or drop leaves that clog gutters.

Clear and check roof. Each fall, clear debris from gutters, downspouts and roof drains. Repair all roof leaks immediately, before they cause more serious damage.

Clear ice properly. Avoid de-icing salts to melt ice on steps and adjacent to the building. Use sand or temporary nonslip treads on steps. Be sure that ice on the building's steps is never removed by hacking with a pickax or spade.

Slope. Ground sloping away from a building may guarantee protection from at least one problem — standing water. (Paul Kennedy)

Remove plant growth. Ivy, moss and lichen should be removed from masonry walls; Virginia creeper can be left, unless the masonry or mortar joints are not completely sound, and should be regularly cut away from the gutters and roof.

Control condensation. Avoid excessively high interior humidity in the wintertime, because it can lead to condensation. Be sure to have at least a little ventilation during winters in the North and summers in the South to reduce condensation.

Repoint loose joints. Joints that have loose, crumbling or missing mortar should be repaired immediately (see Repointing Brick and Stone).

Plant growth. Ivy on a structure can lead to destruction of mortar, thus weakening the wall. (Jack E. Boucher, HABS)

Caulking. This compound prevents moisture from entering and permits movement between two different materials. (Paul Kennedy)

Clean soot. Any heavily soot-encrusted areas should be cleaned (see Cleaning Brick and Stone).

Check caulking on moving joints. Be sure that proper caulking exists in any joints in which movement is expected, such as open joints around windows and doors, in projecting elements such as lintels, parapets and ornaments, and between materials that expand at different rates. Caulking is a resilient (semi-drying or slow-drying) mastic compound, usually of a synthetic composition such as silicone or acrylic, used to seal cracks, fill joints, prevent leaks and, in general, provide weatherproofing and waterproofing. It is worth the small extra cost of using high-quality silicones, polyurethanes and polysulfides. Caulking should not be used as a substitute for mortar in repointing.

Protect projections. Elements that extend from the face of the building can collect water, permitting it to penetrate into the wall. The top surfaces and gaps between elements should be protected with metal flashing. Use corrosion-resistant, nonstaining sheet metal such as lead or lead-coated copper.

CLEANING BRICK AND STONE

Does your building really need to be cleaned? Cleaning is often unnecessary, and more damage can be done to a masonry surface in one day's cleaning with a harmful technique than a century's normal weathering. Thus, the reasons for cleaning must be considered carefully before deciding to clean.

Buildings are usually cleaned for purely aesthetic reasons. Owners, developers or architects may want the building to look new and to have dramatic before-and-after pictures to show. But why should an old building look brand new? Removing all signs of aging sacrifices the character of the building. For example, over time a stone's color may soften and give the building a rich patina that could be destroyed by a radical cleaning technique. Removing this natural surface requires eliminating a portion of the masonry itself and thus accelerates deterioration.

Unless it can be shown that the dirt, paint or stain on the masonry surface is actually harming the masonry, it should be left alone, at least until other more serious problems have been solved. On the other hand, evidence may exist that dirt and pollutants are harming the masonry. An encrustation of soot can absorb moisture and make the surface more vulnerable to atmospheric pollutants and temperature changes. Certain types of stains or surface deposits can also attack the surface of masonry. A reasonable aim, then, is to remove harmful stains, heavy dirt encrustation, dark streaks and sooty grime as much as possible, while leaving the natural characteristics that come with the age of the building.

Most cleaning contractors are able to supply most of the techniques described in this chapter, although many tend to favor those that are the easiest or most profitable for them. The building owner must ensure that the technique chosen is the one that is best for the building; and this is worth paying an extra price, if there is one.

Chemically cleaned granite showing no damage to the stone. (H. Ward Jandl)

Opposite: Steam cleaning Georgia marble cornice of the Corcoran Gallery (1897), Washington, D.C. Steam cleaning was necessary here to avoid bringing iron deposits in the stone to the surface, thus staining the stone. (Paul Kennedy)

New equipment, chemicals and procedures, even ultrasonic descaling and the use of pulsed laser radiation, are being developed; dealing with difficult cleaning situations could be postponed until these methods become commonly available.

HOW TO PLAN A CLEANING PROJECT

In cleaning, as with other rehabilitation tasks, the basic principle is to select the gentlest possible method that will achieve an acceptable level of cleanliness. Cleaning methods generally are divided into three major groups:

Water. These methods soften the dirt and rinse the deposits from the surface.

Chemical. Chemical cleaners react with the dirt and/or masonry to hasten the removal process; the deposits, reaction products and excess chemicals are then rinsed away with water.

Abrasive or mechanical. Abrasive methods include grit blasting (usually sandblasting) and using grinders and sanding disks to remove the dirt by abrasion. All are generally inappropriate ways to clean old masonry.

Water cleaning a brownstone structure. (© Michael Devonshire)

Sandblasting once was so widespread that, to the general public, the word was synonymous with masonry cleaning. Any contractor willing to buy a few thousand dollars' worth of equipment was free to wreak widespread damage on the brick and stone buildings of North America. Today, the problems of sandblasting are commonly known, and there is a shift to chemical cleaning, sometimes even by the same unqualified contractors who used to sandblast; as a result, there is a danger of the same kind of excesses as once occurred with sandblasting.

Often, surface dirt can be cleaned off many kinds of masonry using low-pressure water jets and a light scrubbing with a nonmetallic brush, supplemented if necessary with a nonionic detergent. Sometimes, a simple bristle brush, a pail of water and a garden hose are enough to do the job.

If water washing proves insufficient, a specialist may recommend using an alkaline chemical product on limestone and marble. An acid cleaner is generally used for sandstone, as this material is particularly difficult to clean because of the strong bonds between the dirt and the stone's mineral surfaces. Test

Above left: Federal Hall (1842), New York City. After being sandblasted in the 1960s, the structure was cleaned in the 1970s and covered with a graffiti-proof coating, which attracted water and dirt. In 1986 the coating was stripped with a chemical cleaner and then water rinsed, restoring the marble to its original appearance. (Stanford Golob, ProSoCo, Inc.)

Above right: Sandblasted and unsandblasted brick. (*The Old-House Journal*)

patches will determine whether a particular sandstone is sensitive to acid. Special stains on sandstone can be treated with chemicals by applying poultices (page 105) to the affected area. Limestone should not be cleaned with acidic cleaners because they would stain and dissolve the surface.

There are no easy rules for choosing an appropriate cleaning technique. The choice depends on careful consideration of a number of factors:

- How clean a surface is desired?
- What is the nature of the dirt or soil and how tightly is it adhering to the surface?
- What products can dissolve or detach this soiling?
- What is the type of masonry to be cleaned and what are its characteristics?
- How is the wall constructed (for example, are there any metal attachments that could rust)?
- How can the environment as well as health and safety best be protected?

Deciding how clean a surface should be depends on the type of masonry and the kind of stain or dirt; often one technique can clean most of the building adequately but some heavily soiled areas may require another measure. Also, some masonry is naturally heterogeneous and may have considerable variation in color. The question of evenness also applies to

Conditions that may require cleaning: bird droppings (top, HABS) and staining from metal (above, NTHP).

groups of buildings. If the building is one of a homogeneous row, consider how clean the others are before proceeding; it may look strange to have only one brightly clean building. In some historic districts, permission is given only if the ensemble is to be cleaned at the same time.

Previous treatments of the building and its surroundings, if known, should also be evaluated. Earlier waterproofing applications may make cleaning difficult. Repairs may have been colored to match the soiled building, and cleaning may make these differences apparent. Salts or other snow-removal chemicals used near the building may have dissolved and been absorbed into the masonry, causing potentially serious problems leading to efflorescence and spalling (page 136). Preparations should be made to deal with these problems after cleaning.

Who should do the work? Simple water washing can be carried out by the homeowner or a handyman following the guidelines in this book, but more complex techniques such as chemical and mechanical cleaning are generally done by contractors and should never be undertaken without the advice of an experienced and independent expert.

When not closely supervised, contractors sometimes tend to change methods on site in order to speed up work; in such cases they may use higher concentrations of chemicals or increased pressure of water or abrasive material. Consider insisting that only prediluted chemicals be permitted on site and that the pressure regulators of cleaning equipment be fitted with limiting devices. The surest way to avoid problems is to use reputable and experienced contractors and ensure that they are closely supervised by qualified, independent experts.

What Kind of Dirt Is It?

To remove dirt in the most effective, yet least harmful, manner, determine the dirt's general nature and where it comes from. Soot and smoke, for example, may require a different method of cleaning from oil stains and bird droppings. Other common cleaning problems include metal stains such as rust or copper, graffiti, paint, oil, tar and organic matter such as moss, algae, lichen, fungi and the tendrils left on masonry after the removal of ivy.

GRIME

The most typical form of dirt in urban and industrial areas is, simply, grime. This is a mixture of airborne particles that come from natural and industrial sources and interact with each other and with the byproducts generated by the deterioration of the masonry and micro-organisms, creating a gray or black film or crust on the surface. The dirt also may be a weathered or discolored portion of the masonry itself that cannot be removed without taking away part of the brick or stone.

It is often very difficult to determine the exact nature of dirt even with lab tests; therefore, the most practical way to choose a cleaning method is through test panels on the surface itself. A source of dirt, such as coal soot, may be a past or continuing source of problems; in this case, the underlying problem should be solved before attempting to treat the symptom.

Other possible candidates for cleaning: graffiti (top left, HABS) and biological growth (top right and above, ProSoCo, Inc.).

PAINT

Building owners often wish to strip painted masonry for aesthetic reasons or to avoid regular repainting. Paint is also stripped to solve technical problems: to ensure that a new coat of paint adheres to the surface or to remove a nonporous paint that is trapping moisture.

Top: Grime from industrial pollution. (© Randolph Langenbach)

Above: Staining from atmospheric pollution. (H. Ward Jandl)

Opposite: Water cleaning from a cherry-picker. (© Richard Pieper)

However, paint sometimes plays a protective role on porous masonry, and its removal can hasten deterioration. The process of stripping also can damage fragile masonry such as old bricks. In fact, many old buildings were painted at the time of their construction, or shortly thereafter, to camouflage and protect poor-quality brick, to imitate the color of stone or to hide patches and repairs. In such cases maintaining the paint is not only historically correct, but it also avoids unpleasant surprises.

Before deciding to clean off all the paint, remove paint from several parts of the building to see what the masonry is like underneath, although even this random testing may miss problem areas. If worse comes to worst, the building can always be repainted after cleaning, thoroughly rinsing and possibly neutralizing the surface first to ensure that the new paint adheres.

With buildings of particular historical or architectural significance, the paint type, color and layering should be researched before removing paint. In addition, consider leaving all the original layers of paint on one area of the wall as a record for future owners or researchers.

Choosing a Cleaning Technique

The ease of cleaning depends largely on the nature of the material to be cleaned: its hardness, porosity, surface finish, design and soundness. For instance, it is easiest to remove dirt from smooth, glazed surfaces. Incorrectly chosen cleaning products can cause damaging chemical reactions with the masonry itself, for example, acidic cleaners used on marble and limestone. Other masonry products also are subject to adverse chemical reactions with some cleaning products. Thorough understanding of the physical and chemical properties of the masonry can help avoid the inadvertent selection of damaging cleaning materials.

Other building materials also may be affected by the cleaning process. Some chemicals may have a corrosive effect on paint or glass. The portions of building elements most vulnerable to deterioration may not be visible, such as embedded ends of iron window bars. Other totally unseen items, such as iron clamps or ties that hold the masonry to the structural frame, also may be subject to corrosion from chemicals or even plain water. The only way to prevent problems in these cases is to study the building construction in detail and to evaluate cleaning methods with this information in mind.

Testing Cleaning Methods

The selection of a cleaning technique is best done by testing several methods, beginning with the gentlest, to determine an acceptable level of cleanliness. Homeowners should seek the assistance of an outside expert in conducting these tests.

Because a single building may have several types of masonry materials and each material may have a different surface finish, each of these different areas should be tested separately. There also may be several types of stains or paint to be removed, and no one method may work for all. The results of the test may well indicate that several methods of cleaning should be used on a single building.

The tests should begin with water washing — the gentlest technique — and move up to harsher methods only if the simpler ones do not produce a satisfactory result. The tests should include several intensities with each method; for example, with water washing, try different pressures and durations; with chemicals, test various concentrations.

Apply cleaning tests to an area of sufficient size to give a true indication of effectiveness. Test patches should encompass at least 2 square feet; with large stones, a test should include several stones and mortar joints. Patches should generally be located in a less visible part of the building and should include each different element to be cleaned.

To judge how effective the cleaning operation was, let the masonry dry thoroughly and carefully examine the results, looking for surface deterioration or discoloration. It is best to allow test patches to weather for at least a few weeks to make sure that stains or efflorescence do not appear, particularly with chemical cleaning. For a building of considerable historical significance, the testing should be done well in advance — a year is not an unreasonable length of time — to expose the masonry to a full range of seasons. Any delay is unimportant compared to the potential damage and disfigurement that

Which Cleaning Product to Use?

Here are some products that may be used to clean particular stains or types of dirt.

Graffiti, spray paint: acetone, varnish, solvent, commercial paint strippers

Asphalt, bitumen, tar: kerosene, xylene, toluene, mineral spirits, chlorinated solvents, automobile asphalt cleaner

Oil, grease: chlorinated solvents, petroleum solvents, ammonium hydroxide

Rust: oxalic acid, sodium citrate (for carbonate stones), citric acid

Copper and alloys: ammonium carbonate, ammonia solutions, sulfamic acid, ammonium chloride

Lichen, moss, ivy, various plants: commercial herbicides, detergents, household bleach

Algae, mushrooms, mold: soap, commercial powdered cleanser with bleach, chlorinated lime, leaching powders, peroxide

may arise from use of an incompletely tested method.

The cleaning budget should include money to pay for these tests. Usually contractors are more willing to conduct a variety of tests if they are reimbursed for their time and materials, particularly if the tests include methods with which the contractors are not familiar.

Environmental, Health and Safety Concerns

The potential environmental effects and health dangers of each proposed cleaning method should be carefully evaluated.

Chemical cleaners in both liquid and vapor forms can cause serious injury to cleaning operators and passersby. Even in dilution when washed off with water, chemical cleaners may damage trees, shrubs, grass and plants. Animal life, ranging from domestic pets to songbirds to earthworms, also may be affected by the runoff.

Mechanical methods of removing dirt create airborne dust that can pose a serious health hazard, particularly if poisonous lead-based paints are being removed or if the abrasive material or masonry contains crystalline silica. Common sand contains this mineral, which causes silicosis, a serious lung disease; for this reason, nonsiliceous abrasives, such as crushed limestone, are preferred if sandblasting is ever used. If siliceous sand is used or if the material being cleaned contains silica, such as sandstone, a wet-grit method must be used to help keep down the dust.

The proposed cleaning project also may cause property damage. Wind, for example, may blow cleaning chemicals onto nearby automobiles, causing etching of the glass or spotting of the paint finish. Similarly, airborne dust can enter surrounding buildings and excess water can collect in nearby yards and basements.

Equipment operators should wear protective clothing (rubber gloves, suits and face protection); with dry-grit blasting, they should also wear helmets that provide filtered air. All other workers on the job site who are exposed continually also must be protected. Scaffolding should be enclosed to protect the public. Adjacent parts of the building could be covered with plastic. Also, protect vegetation with plastic; presoak lawns and other vegetation and mist with clean water during the cleaning process. With acid cleaning, which will etch any exposed glass, valuable and vulnerable elements, such as stained glass and glazed ceramic floor tiles, should be covered with strippable latex rubber (spray or liquid) or a combination of plastic sheets (15-millimeter polyethylene sheeting with caulking at the edges) and close-fitting boards for added protection.

WATER CLEANING

Cleaning masonry with water is generally the simplest operation, the safest for the building and the environment and the least expensive; unfortunately, it is often the least considered method.

The four most commonly used water-cleaning methods, from the simplest to the most complicated, are handscrubbing, spraying, pressure washing and steaming. Each of these methods can be used on its own or in combination with other techniques such as a finishing rinse with chemical products. They are particularly effective on calcareous (calcium-based) materials such as limestone and marble and on some lime-based bricks where the dirt has adhered to the masonry by gypsum, a material that is partially water soluble.

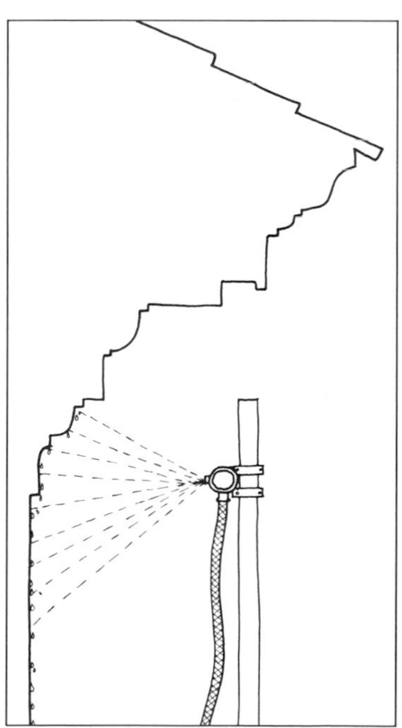

Left: Preparation for water cleaning. Windows and doors are sealed with plastic. (Dinu Bumbaru)

Right: Water washing over a period of time. Hoses are installed above the area to be cleaned, and water is allowed to drip gently over the surface, loosening dirt. (Christina Henry)

Handscrubbing

Simple handscrubbing is a surprisingly effective cleaning method in many situations. All that is needed is a bucket, a bristle brush and a garden hose. A soft natural-bristle, nylon or fiber brush is preferred; steel brushes are too abrasive and leave specks that rust. Wet an area by sprinkling with a hose, then brush it (using a little extra elbow grease for the dirtiest areas) and, finally, rinse the area with the hose. A small amount of a liquid nonionic detergent such as dishwashing liquid can be added to the water. Handscrubbing is somewhat time consuming and is most effective with weakly adhering dirt; it is most often used on glazed and polished surfaces, on buildings of particular significance and by homeowners doing their own work.

Spraying

Spraying relies only on a hose using regular water pressure. A fine mist of water for several hours loosens the bond of the dirt, which is then washed off. In general, the finer the mist, the better. The spraying process should be interrupted regularly to reduce the quantity of water used and to allow the soiling encrustation to swell and contract, which helps loosen it from the masonry. A timer can be installed on the hose for this purpose: for example, it can be set to stop for four minutes after every 10 seconds of spraying, making sure not to let the masonry dry completely between spraying. Afterward, a rinse with a garden hose, and with additional manual scrubbing if necessary, should be enough to remove the loosened dirt from the wall.

To set up the spraying operation on a small building, the hoses and heads are usually mounted at the top of the wall on scaffolding. For a large building they are moved periodically down the wall in order to cover the entire surface area; extra heads can be placed at heavily soiled areas and to reach hard-to-get-at places. Three commonly used misting techniques are

- Dripping water down the wall through a perforated hose placed at the top
- Spraying the wall with a low-volume, wide-angle sprinkler (1/3 gallon per minute)

- Projecting a fine mist on the wall by covering a hose head with a cloth

When using a technique requiring large quantities of water, check the pH of the water as well as the presence of trace metals, especially iron. Acidic water, which can damage certain limestones and marble, should be treated before use; water containing iron compounds can stain masonry.

Pressure Washing

Pressure washing depends on mechanized pressure. Low-pressure water jets (about 500 psi) can be used in certain circumstances to clean stone or hard brick although they may damage very soft materials. This method can be effective alone on loosely adhered dirt and is most often used in association with chemicals. Higher-pressure jets, in fact, have the same effect as abrasive cleaning, and the use of such pressures should be avoided on old masonry buildings. It is important to keep the hose nozzle moving, and special care must be taken on soft lime mortars, which are very easily removed.

Steam Cleaning

Although once a common method, steam is now rarely used because it is slow, generally no more effective than plain water and poses safety problems for the operator. It may still

Steam cleaning a downtown commercial building. (Tom Lutz)

be useful, however, in removing oily stains, such as those from chewing gum, and dirt on highly carved or ornamental surfaces with a low risk of surface abrasion. Steam is generated in a flash boiler and directed against the masonry surface at a low pressure of about 10 to 30 psi using a nozzle with a ½-inch aperture. Detergents and chemicals may be added to supplement the cleaning power of the steam; caution must be exercised because some combinations may be hazardous to the operator or masonry.

Damage to sandstone from too high a water pressure. (Michael F. Lynch)

Problems with Water Cleaning

Water cleaning can present a few problems. Water can seep through deficient seals around windows and doors. Also, porous masonry may absorb excess amounts of water during the cleaning process, particularly if large quantities are used. The excess water can penetrate through to the interior of a building, causing water stains and mold. Within the wall, it can damage insulation and corrode hidden metal elements, leading not only to rust stains, but also to a weakening of structural supports. With a well-sealed wall and a technique that uses only a limited amount of water, the degree of water penetration, however, is tolerable even with moderately absorbent masonry walls.

If a wall is washed repeatedly, excess water also can bring soluble salts from within the masonry to the surface, forming efflorescence; in dry climates, the water may evaporate inside the masonry, leaving the salts slightly in back of the surface. Efflorescence, while not in itself harmful to masonry, does provide a warning sign that moisture is present.

Water used for cleaning may mobilize other impurities in stone, such as metallic ions that can cause permanent rustlike discoloration. A similar effect is caused by impurities such as iron and copper in the water supply itself. The chlorides in chlorinated water can bring iron to the surface. Even "soft" water may contain deleterious amounts of these chemicals. To limit iron and rust in the water, it is best to use plastic rather than ferrous hose fittings.

Water methods must not be used during periods of cold weather because water within the masonry can freeze, causing the stone or brick to spall or split and crack. Because a wall may take weeks of good weather to dry after cleaning, buildings should not be water cleaned within a few weeks of the first average frost date or whenever local forecasts predict cold weather.

The following precautions thus should be taken when water washing:

- Ensure that the wall is watertight, all damaged areas have been repaired, and masonry and joints, including mortar and caulking, are sound.
- Use techniques that require the least amount of water.
- Use plastic or nonferrous tools.
- Seal all door and window openings with a 15-millimeter polyethylene sheet taped all around; a sheet of plywood can be placed in front of the polyethylene for additional protection.
- Allow a few weeks for adequate drying of the masonry to ensure that the wall will dry out before frost.

Chemical Cleaning

The use of chemical products in masonry cleaning has greatly increased in the past few years. Chemical cleaning is generally an acceptable technique for cleaning old masonry and is certainly the most effective and least damaging method of removing paint. If not properly carried out, however, it can also cause staining, efflorescence or other damage; it is also potentially very dangerous to the operator, and extreme caution must be exercised.

The main types of chemicals used are acids, alkalis and organic compounds. Chemicals discussed in this section are those most often used for removing common soot. Once the dirt is removed, bits of tar, spots of oil, rust and all kinds of other stains on the masonry may be revealed. While this book is not an exhaustive compendium of every cleaning chemical and technique, the table on page 95 gives a general idea of what cleaning procedure best remedies particular dirt removal problems. Other resources are listed in Further Reading, particularly the *Masonry Conservation and Cleaning Information Kit* developed by the Association for Preservation Technology for professionals.

Testing cleaning methods. Brushing brick with an acid cleaner (top) and removing it with water (center). An alkaline cleaner on limestone (above), used to remove paint, works for a few hours before removal. (Dinu Bumbaru)

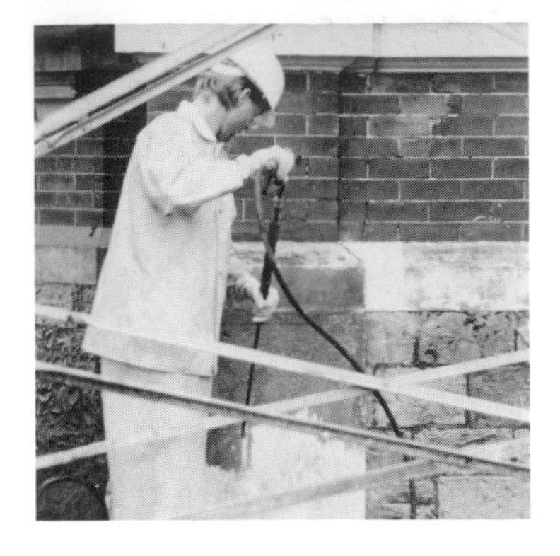

Testing cleaning methods. Spraying off the alkaline cleaner (top) and neutralizing the surface (center). A cleaner is applied underneath a sill (bottom). (Dinu Bumbaru)

Techniques

Some variation of the following method generally is used in chemical cleaning:

- The wall is first sprayed with water. Wetting it avoids excessive penetration of chemicals and soil into the masonry and also helps soften the soiled crust. Paint stripper, however, is applied to dry masonry.
- The chemical products are then spread over the surface of the masonry with a brush or a sprayer. These products include surfactants — organic compounds with powerful properties of detergency and wetting to clean and penetrate masonry.
- Finally, the cleaners and dissolved soils are removed by thoroughly rinsing with a low-pressure jet of warm water, generally less than 500 psi (page 99). With some alkaline cleaners, the use of a neutralizer before the final water rinse is advised.

Another method, used for special problems in limited areas, involves applying the chemical in a paste called a poultice (page 105), after crusty soil has been removed. Sticky products such as chewing gum, adhesives or tar can be hardened with dry ice and then broken off.

Acidic Cleaners

These products are used for cleaning most granites, sandstones and all noncalcareous stones and unglazed bricks. The acid loosens the dirt, which can then be rinsed off easily. To determine if the stone is acid soluble, put a few drops of diluted hydrochloric (muriatic, HCl) acid on the surface; if it bubbles or foams, the stone should not be cleaned with acid.

Hydrofluoric (HF) acid is the most commonly used acid cleaner, as it is the only one that does not form soluble salts that harm brick and stone. However, it is very toxic and corrosive and will damage glass and aluminum, which must, therefore, be protected. It should be used in a maximum concentration of 5 percent although it is preferable to use a weaker solution of 1 to 3 percent plus a second application on heavily soiled areas. Because acids can lead to rust stains on certain stones by dissolving otherwise insoluble iron compounds, some include rust inhibitors. "Complexing agents" such as phosphoric acid (H_3PO_4) or EDTA are sometimes added to acid in a limited concentration to reduce salt deposits.

Hydrochloric acid is strong but less efficient as a cleaner and should not be used on old masonry; it causes efflorescence and color changes if the stone has a high iron concentration. Old brick buildings are particularly susceptible to damage from this form of acid.

Acid should never be used to clean limestone, marble, sandstone that contains calcium carbonates and most polished surfaces; it will cause serious damage to these stones and their finishes.

Alkaline Cleaners

Alkalis (caustics) can be used on acid-sensitive masonry materials such as limestone or marble, glazed brick and glazed terra cotta, although they are rarely needed because most of these materials respond well to cleaning with water and detergents. The most commonly used products have a potassium hydroxide, ammonium hydroxide (ammonia) or sodium hydroxide (caustic soda) base. Sodium hydroxide requires greater effort to rinse thoroughly and should not be used on weak, friable or very old and historic masonry, because it tends to cause salt deposits.

Alkaline cleaners can produce efflorescence and may result in brownish stains if iron compounds exist in the stone. As a safeguard, the surface to be cleaned should be wetted beforehand and neutralized afterward by rinsing with a mild acidic solution such as acetic acid. To avoid stains, alkalis, like acids, must not be used on stones with very high iron content. Alkalis can also damage paint, wood and some metals.

Paint Removal

Limited removal of excess layers of badly peeling paint in preparation for repainting should be carried out by hand using natural-bristle or nylon brushes and handscrapers.

A heat gun (with a temperature of less than 750°F) may prove useful for paint removal in some cases; however, propane torches or any other type of open flame should never be

Revealing earlier repairs. A test patch of brick (top and center) reveals an area on the left that had been made to look like brick. Cleaning has removed the area's original bricklike appearance. The patch is then rinsed (above). (Dinu Bumbaru)

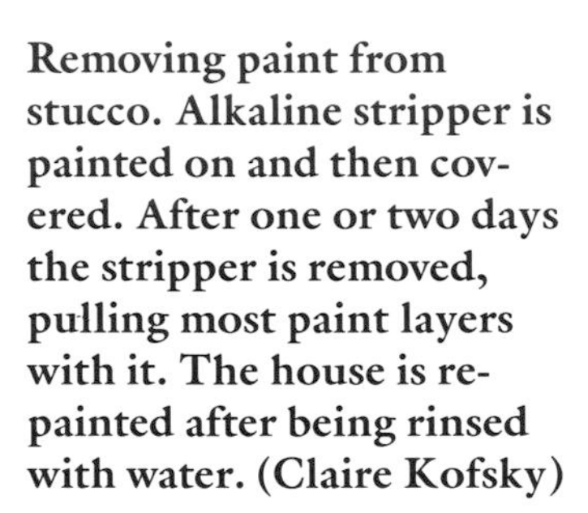

Removing paint from stucco. Alkaline stripper is painted on and then covered. After one or two days the stripper is removed, pulling most paint layers with it. The house is repainted after being rinsed with water. (Claire Kofsky)

used on an old building; they not only can damage masonry and other building materials, but they also present an extremely high risk of starting a fire.

In general, the only technique that will completely and effectively remove paint is the application of chemical paint removers, generally alkalis or organic solvents. Most commercial paint strippers also contain a thickening agent or gel that enables the compound to cling to a vertical surface. The length of time the remover must remain on the surface can be up to several hours; the work area should be kept moist by adding chemicals as required.

Alkaline strippers with a base of sodium or potassium hydroxide are generally more economical and work particularly well on old linseed oil–based paints. They can cause efflorescence, however, and, like all cleaners, must be thoroughly rinsed off. A weak hydrofluoric acid–based cleaner is sometimes applied as a neutralizer before the water rinse.

Organic solvents such as methylene chloride tend to be more expensive, but they are more effective at dissolving most types of paint as well as more recent products such as urethane varnishes or epoxy. They can spread stains deeper into the masonry, however, unless used in poultice form — not always a practical way to remove paint from a large area. They are generally quite toxic and used only when other methods are not adequate.

Some other coatings used as paints, such as lime washes (including whitewash and color wash), are soluble in acid. If there are several layers of paint on a wall, it may be necessary to use different solvents on different layers.

Graffiti removal presents another problem. If the wall previously was dirty, the cleaned area might be lighter than the rest of the wall and the ghost of the offensive message still evident. The whole wall must be cleaned, or an attempt can be made to artificially dirty the cleaned area so that it matches the rest of the wall.

POULTICES

Poultices, or leaching packs, involve spreading a chemical paste on the masonry to loosen the dirt or stain. The packs then draw out the dirt as they dry, thus avoiding the reabsorption of dissolved material into the masonry. This technique is used to remove stains from porous masonry and for more general cleaning problems inside buildings where it is impossible to use water.

Poultices may be applied successfully to remove stains such as oil, tar, plant materials (lichens and algae), graffiti (including spray paint), metallic stains such as iron and copper and occasionally some types of salt deposits (efflorescence).

The paste is made up of an inert, absorbent material (such as talc, chalk powder, clay, sawdust, fuller's earth, diatomaceous earth, whiting or even shredded paper) that has been saturated with a solvent to dissolve the specific type of stain. Obtain these inert materials in pure form, ensuring that they are not, in fact, acidic. Glycerine is often added as a thickener and to slow down evaporation.

Generally, a layer about ⅓ to ½ inch thick is applied on prewetted masonry. Poultices should be applied to whole stones up to the joints to avoid patchy-looking, unevenly cleaned stones. The entire area is then covered with a plastic sheet to slow down the drying process. Once the poultice has dried out, it can be removed carefully by hand with the aid of wooden spatulas and bristle brushes. Finally, the area is rinsed off with water.

Poultice. Absorbent material is applied to a stain; the area is covered; the poultice is scraped off with a wooden spatula and the surface is rinsed with water.

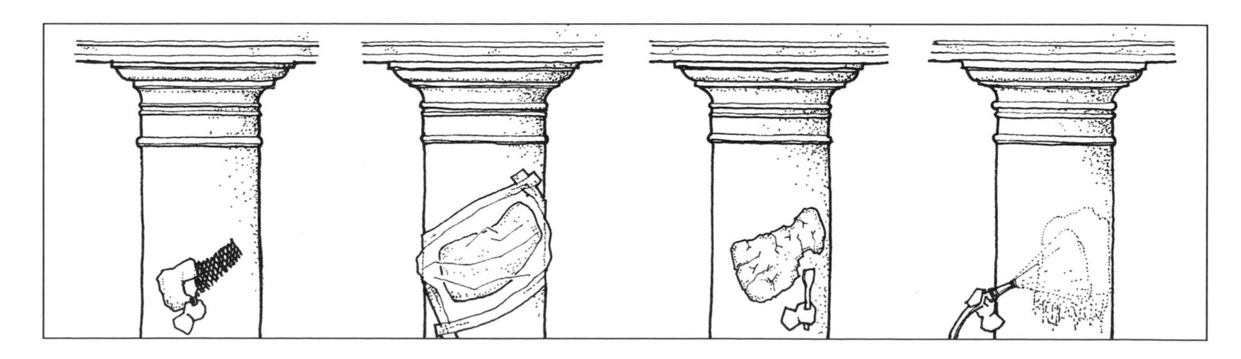

Problems with Chemical Cleaning

If chemicals are not used properly, they can cause water damage, dissolve certain types of masonry and produce discoloration. Chemicals can also be very dangerous to users' health and to the environment.

Because most chemical cleaners are water based, they have many of the potential problems of plain water. However, compared to water washing, the use of chemicals somewhat reduces the quantity of water needed, limiting the risk of water penetration.

Incorrectly chosen cleaning products can cause damaging chemical reactions with the masonry itself. Marble and limestone, for example, are dissolved easily by acidic cleaners, even in dilute forms. Chemicals also may change the color of the masonry rather than remove soil and may leave a hazy residue

Sandblasting paint on a brick wall, a method that is not recommended. Such surfaces often must be repainted to protect the damaged brick. (Ward Bucher)

in spite of heavy rinsing. In addition, chemicals can react with components of mortar, stone and brick to create soluble salts that can form efflorescence. But advance testing should help avoid these problems.

Chemical cleaning should be done only by professionals. Unfortunately, too many contractors have little expertise or understanding of masonry, different types of staining or even the chemicals with which they are working. When approaching cleaning companies, request references and look at projects they have completed. If chemical cleaning is determined to be the best course of action, ask what chemicals will be used and insist on proprietary cleaning products that have been properly tested and are backed by the reputation of a known manufacturer. Avoid the secret formulas that cleaning contractors mix up in their garages; if these truly are old formulas, they probably are based on hydrochloric acid; it is a good idea to prohibit use of any hydrochloric acid. These formulas also may well be pure hydrofluoric acid without other useful components.

Abrasive Cleaning

Abrasive or mechanical cleaning techniques — such as sandblasting and the use of power sanders — are unacceptable cleaning methods for old and historic masonry. These techniques destroy the appearance, original materials and physical well-being of a building. Abrasive cleaning of masonry causes the following problems:

- Damage to and often destruction of decorative detailing or texture
- Removal of the hard-fired exterior surface of brick or terra cotta or the hard skin of stone, thus exposing the softer interior to weathering and rapid deterioration
- Destruction of pointing and joint details, leading to water penetration
- Roughening and pitting of the surface, increasing the possibility of water and dirt accumulation and causing rapid and uneven soiling as the masonry ages
- Damage to other parts of the building because abrasive techniques are hard to control

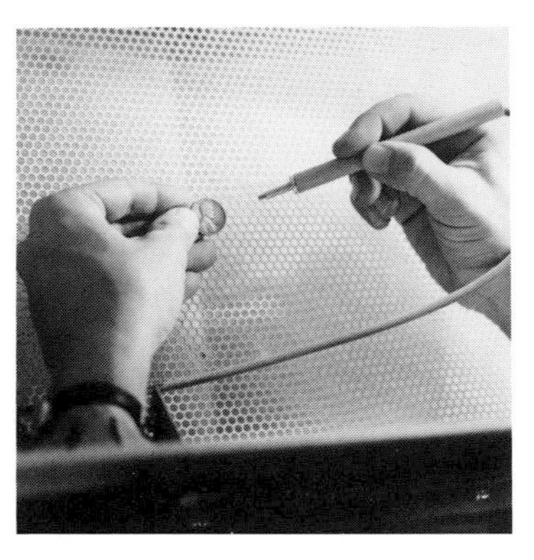

Top: Damage from wet-grit blasting. (Lee Nelson)

Above: Microabrasive cleaning using a small pressure instrument. This technique has limited applicability because it is expensive and requires a highly trained user. (National Park Service)

- Environmental damage by spreading dust or slurry, in the case of wet-grit methods

Mechanical cleaning using tools such as wire brushes, rotary wheels, power-sanding disks and belt sanders does fundamental damage. It

- Removes masonry surfaces
- Causes accidental scour marks and other damage
- Deposits small metallic particles that can cause rust, thus staining the masonry

Blasting

Blasting is the most common abrasive technique and involves spraying under pressure a material that impacts or abrades the surface: sand, ground slag, crushed walnut or almond shells or rice husks, ground corncobs or coconut shells, crushed eggshells, silica flour, synthetic particles, glass beads and microballoons. Wet-grit or hydrosilica cleaning combines a grit material such as those listed above with water. The detrimental effect on masonry is the same with this method, however.

While not normally considered an abrasive cleaning technique, water at too high a pressure can also cause masonry surfaces to abrade. On the other hand, adding a very small and carefully regulated amount of grit to a carefully controlled, pressurized water stream may not cause the problems normally associated with abrasive cleaning.

A delicate method of abrasive cleaning being adopted by architectural conservators uses a microabrasive grit on small, hard-to-clean areas of carved, cut or molded ornaments on building facades. Originally developed by museum conservators for cleaning sculpture, this technique employs a very fine powder such as glass beads, microballoons, aluminum oxide, silica flour or crushed dolomite powered at approximately 35–50 psi by a very small, almost pencil-like pressure instrument. This technique has limited practical applicability on a large-scale building cleaning project because of the cost and the relatively few persons competent to handle the task. In Europe sophisticated variations of microabrasive cleaning and low-pressure water techniques have been developed and are beginning to be used on this side of the Atlantic.

Mitigating the Effects of Abrasive Cleaning

What can be done if a masonry building has already been sandblasted? Unfortunately, not much.

It is impossible to recover lost crisp edges, tool marks or other indications of craft technique. Harder, denser stone of uniform composition, however, should continue to weather with little additional deterioration. Unfortunately, some types of sandstone, marble and limestone will weather at an accelerated rate once their protective surface has been removed.

Softer types of masonry and particularly brick are most likely to require some remedial treatment if they have been abrasively cleaned. Because the potential treatments are both technically and aesthetically problematic, it may be best not to intervene unless there are definite signs of deterioration, such as spalling.

Brickwork whose hard outer surface has been removed is much more susceptible to spalling as a result of moisture and frost action in winter months. Therefore, to protect these abrasively cleaned structures it is especially important to eliminate sources of excessive moisture (defective flashing and gutters, sills with no drips, water splashed up from the ground, for instance). Selective repointing with weak, more porous mortar might encourage masonry to dry out.

Only if evidence of spalling is apparent should more general action be taken. This could include painting or applying a good, breathable surface treatment (page 138) that may slow deterioration. It also may be necessary to replace some bricks. If a brick surface has been so extensively damaged and spalling is widespread, drastic action is required. In buildings of particular importance, where conservation of the original historic building materials is considered of paramount importance, the wall could be covered with stucco, if it will adhere, or another material (page 144); of course, the application of this treatment means that the original appearance will be sacrificed. However, keep in mind that the original color and texture was changed already by the abrasive treatment and, at this point, it is more important to preserve the brick itself. As a last resort in the case of severely spalling brick, the masonry itself can be replaced — a difficult, lengthy and expensive process, particularly if custom-made reproduction brick must be used.

Waterproof coating on sandblasted brick. When the sealer was applied, it failed to dry properly because of cold temperatures and dripped, making the surface cloudy. (National Park Service)

Repointing Brick and Stone

Repointing is the process of removing deteriorated mortar from the joints of a masonry wall and replacing it with new mortar. Properly done, repointing (also called, somewhat incorrectly, tuckpointing) restores the visual and physical integrity of the masonry. Improperly done, repointing not only detracts from the appearance of the building, but may in fact cause physical damage to the masonry units themselves.

A wall's mortar joints bind together the individual masonry elements into a structural whole ensuring a watertight seal. The bed of mortar also compensates for irregularities in the stones or bricks, which would otherwise lead to uneven stresses and cracking of the masonry units; the more regular the stone or brick, the thinner the joint can be.

A wall made up of many small units such as brick or stone is both easy to construct and absorbs inevitable slight movements, including variations in temperature, settlement of the building and vibrations. To absorb these movements, the mortar joints must be somewhat weaker than the masonry units; otherwise, the masonry units become the weakest part of the wall, and slight movements would cause the brick or stone to crack or spall. If mortars are too strong, they tend to be more impermeable to moisture than the masonry units and thus prevent drying through the joints; moisture movement then is concentrated in the brick or stone, leading to damage of the masonry.

Mortar does not play a structural role in some masonry. Some rural garden and field walls are "dry" stone walls, meaning that no mortar is used at all. And in some very old masonry buildings and the foundations of more recent small farm buildings, stones were laid directly on each other, after which mortar was added only to help keep weather out of the joints.

Deteriorated brick wall. Too hard a mortar was used with this soft brick. (David Baxter)

Opposite: Former V.C. Morris Store (1948), San Francisco, designed by Frank Lloyd Wright. The architect's attention to detail often extended to mortar, which he often had tinted two colors to differentiate vertical and horizontal bonds, thus creating an almost seamless horizontal look. (Carleton Knight III)

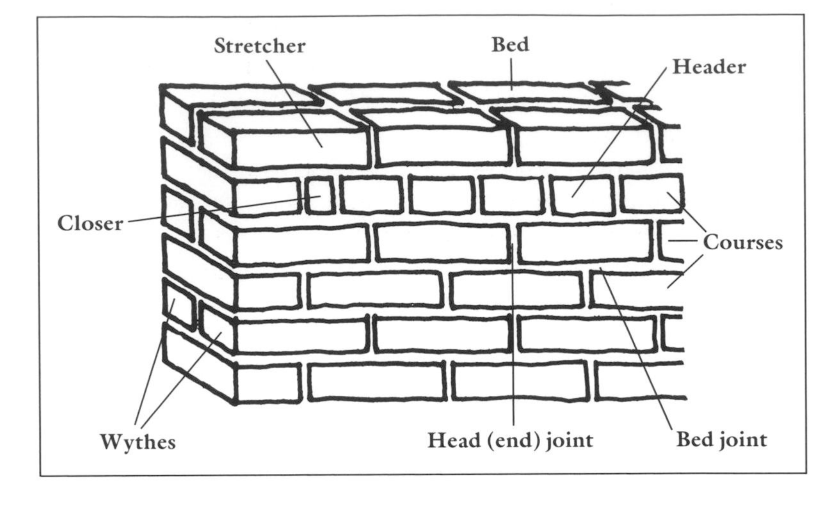

Section of a brick wall. (Robert D. Loversidge, Jr., AIA, from the *Old-Building Owner's Manual,* Ohio Historical Society, by Judith L. Kitchen, 1983)

Unlike most other parts of a building, mortar joints are not designed to be permanent, although a good pointing job should last from 50 to 100 years. When the time comes to repoint, shortcuts and poor craftsmanship will only result in a job that needs to be done again soon. Repointing is not a job for an inexperienced do-it-yourselfer.

Planning a Repointing Project

Repointing is a lot harder than it looks. Preparing the joints properly, getting a good color and texture match, and filling the joint so that the mortar will adhere properly is the job of an expert.

Repointing will probably be both expensive and time consuming because of the extent of handwork and special materials required; however, repointing only those areas that require work rather than an entire wall (as is often specified by architects or suggested by contractors) may avoid unnecessary expense.

To prevent freezing or excessive evaporation of the water in the mortar, repointing is best carried out when wall temperatures are between 40° and 95°F. During hot weather, repointing should be done on the shady side of the building to slow the setting; the wall should be covered with burlap or a tarpaulin while the mortar sets.

It is generally better to remove paint or clean the masonry before repointing. However, if the mortar is badly eroded, allowing moisture to penetrate deeply into the wall, the joints should be temporarily plugged (for example, with foam backer rods or strippable caulking), or repointing should be done before cleaning.

Repointing should be carried out as one operation after all masonry repairs are completed. Structural or roof repairs should be scheduled so that they do not interfere with repointing and so that all work can take maximum advantage of scaffolding.

Is Repointing Necessary?

The decision to repoint is most often related to some obvious sign of deterioration (disintegrating mortar, cracks in mortar joints, loose bricks, damp walls or damaged plasterwork). Because repointing alone will not solve all these problems, their true cause should be determined and dealt with before repointing, or the mortar deterioration will continue and the repointing will have been a waste of time and money.

One of the main causes of mortar deterioration is weathering, i.e., erosion due to wind and rain (particularly acid rain on lime mortars). This can be compounded by problems of building design, construction or maintenance. For example, a roof shape that allows rainwater to drain onto one part of the wall coupled with an absence or deterioration of gutters can hasten mortar damage. Mortar joint shapes that collect rather than shed water further aggravate the problem. Once water penetrates into small cracks in the joints, it further deterio-

Brick deterioration. Water has gotten behind this brick wall, causing the mortar to disintegrate and bricks to break apart. Because the original cause of the problem was not corrected, repointing was unsuccessful; this can be seen in the hairline crack in the new mortar. (Paul Kennedy)

rates the mortar, particularly in northern climates where the water freezes and exerts a powerful expansion force.

Other factors leading to deterioration are building settlement and the differential expansion and contraction of the mortar and the masonry because of daily and seasonal temperature changes. A common cause of problems, particularly today, is poor bonding between the mortar and masonry because the former was poorly mixed, the joint was not cleaned out well or because an excessively hard mortar has shrunk.

As a general rule, if the pointing is firm, intact and not eroded more than ⅓ inch, it should be left as is. To judge whether joints need repointing, use the following criteria:

- Open joints: the mortar is deeply eroded (more than ⅓ inch from the face of the masonry) or has fallen out.
- Cracked joints: cracks (hairline width or larger) have formed in the mortar.
- Separated joints: the mortar and the masonry do not adhere, resulting in a crack or gap between the two, or the mortar is sitting loosely in the joint.

It is unlikely that all the joints of a building will need repointing. Just because a screwdriver can be poked into the mortar joint does not necessarily mean it needs replacing; the lime mortars found in most old buildings should not be very hard. Good practice calls for repointing only the part of the wall that really needs it, not necessarily the whole wall. However, to maintain a uniform appearance, some sound mortar may be sacrificed if a majority of mortar joints in the same location have deteriorated and need to be replaced.

Carefully assess whether fine joints in stone or brick need repointing, as removing old mortar here presents an especially difficult task. Also, removing portland cement mortar is quite difficult, and the removal process could prove more harmful to the masonry than leaving the mortar in place, unless signs of spalling are obvious.

Test Panels

In choosing contractors or masons, it is useful to have them prepare a test panel on which joint preparation and repointing skills can be demonstrated. First, select an area of masonry in good shape to serve as the control section against which to compare the test panel. Then, choose an area, usually a three-foot-square section of masonry, preferably in an inconspicuous location that includes the types of masonry, joint styles and variety of problems expected on the actual job, and have it repointed. The joints of the panel should be checked before being filled to ensure that they are properly cleaned out and checked again on completion against the original control panel. Once accepted, the test panel can serve as a standard reference for the entire job.

MORTAR INGREDIENTS

After deciding to repoint, the next question is what kind of mortar to use. All mortars are made from the same basic ingredients: water, an aggregate (usually sand) and a binder (usually lime and cement). Although clay was used briefly in the early days of colonial settlement, lime was used almost exclusively as a binder until the introduction of much harder portland cement in the 1870s. Additives are sometimes used to modify the color, increase resistance to frost and speed up the setting time of mortar, although their use is not generally recommended for old and historic buildings.

Modern materials specified for repointing mortar should conform to specifications of the American Society for Testing Materials (ASTM) or the Canadian Standards Association (CSA).

Aggregates

Aggregates make up the largest component of mortar. While sand is now used almost exclusively, other products also served as aggregates in old buildings.

Sand gives mortar most of its characteristic color and texture. Historic sand colors may range from white to gray to yellow within a single sample. Also, because sand often was not screened and graded as it is today, the size of grains may vary from fine to coarse. Therefore, in order to match the range of colors and grain sizes in the original sample, it may be necessary to obtain sand from several sources and then combine it. Even in a simple, small job, some coarse-grade sand may have to be added to standard packaged sand, unless joints are so fine that the look of the mortar does not play an important visual role.

Natural beach or river sand has rounded edges, as seen under a magnifying glass or low-power microscope. Natural sand provides a better visual match with old mortar and can produce good plasticity with less water, allowing the mortar to be forced into the joint more easily and forming a better contact with the old mortar and the masonry. Manufactured sand, made by crushing stone, has sharp, angular edges; it generally produces higher strength mortar but may be more difficult to work and to match visually with old mortars.

Specify that sand in repointing mortar be clean and that it match the original as closely as possible to provide the proper color without other additives. Samples should be approved before work starts. Sand should conform to ASTM C-144.

Other aggregates in mortar usually make up a very small proportion of the total. These aggregates, however, can be important in achieving a good color and texture match. For historic buildings, it is especially important to identify them in the original mortar and specify them in the new one, suggesting if possible where they may be obtained. Other materials that may be found in old mortars include animal hair, clay particles and partially burned lime.

Binders

As the word implies, binders in mortar are a cementing material generally made up of lime and portland cement in various proportions depending on the masonry job.

Lime is a binder of choice for repointing old masonry. It can be used alone, or, in certain cases, cement may be added to improve setting time and durability. The proportion of cement may be increased according to the strength of the masonry and the exposure of the wall.

High-lime mortar is soft and porous and changes little with temperature fluctuations. Because it is slightly water soluble, it can reseal hairline cracks by combining with moisture in the air.

Until World War II, quicklime (calcium oxide) was used, made by burning limestone (calcium carbonate). It was then slaked by adding water to create hydrated lime (calcium hydrate); lumps of fresh quicklime were added to water, the mixture was stirred until the chemical reaction was complete, it was strained through a mesh and, finally, stored for at least two weeks under water in sealed containers. Handling lime in this form was quite dangerous, even for experienced workers.

Since World War II, prehydrated lime — a powder form, which can be mixed immediately before use and with more safety — has generally replaced quicklime, despite the fact that traditionally slaked lime is said to produce a better-quality mortar. Hydrated lime should conform to ASTM C-207, Type S, Hydrated Lime for Masonry Purposes, which is designed to ensure high plasticity and water retention without being too strong.

Cement was widely used by the end of the 19th century; the most common cement powder in use today is portland cement, an extremely hard cement that is impermeable to water. Much too hard to be used as the only binder in mortar, particularly for old walls of soft brick and stone, ordinary gray portland cement may also have a high content of harmful soluble salts. However, white, nonstaining portland cements have a low alkali content, which can help avoid efflorescence, and are usually mixed with lime to quicken setting. Speeding the initial set can improve durability and increase frost resistance.

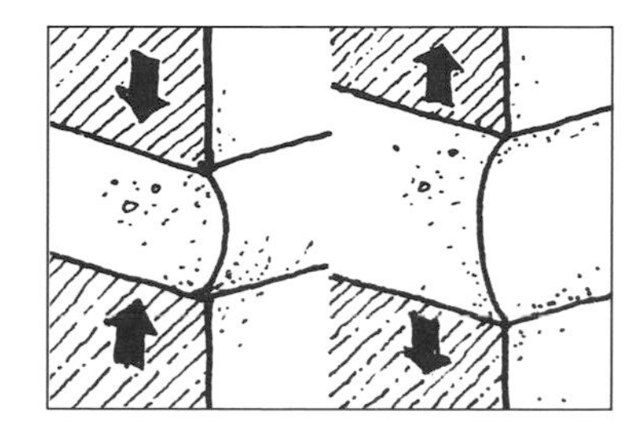

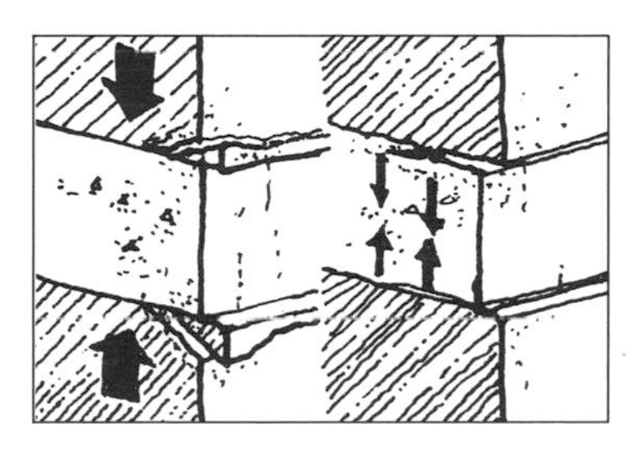

Top: Soft lime-based mortar. It retains its bond with the masonry despite changes in temperature.

Above: Mortar with a high percentage of cement. It becomes too rigid and shrinks more when drying, leaving gaps between the mortar and masonry surface.

Portland cement combined with gypsum, hydrated lime or limestone dust, as well as clays and other substances, is sold as premixed masonry cement in a variety of colors. In northern areas, this mix includes agents to increase frost resistance.

There are two problems with masonry cement. It is generally too hard, particularly with the soft bricks and weak or fractured stone of old historic buildings. Adding more than the normal amount of sand might weaken it enough to make it acceptable for use with harder brick and stone. A second problem is that the composition of the mix, rarely indicated on the package, can vary considerably from one manufacturer to another. If an acceptable mix is found, masonry cement can save time and is less subject to inconsistency than mortars mixed on site; however, building owners should specify that contractors not substitute another brand.

Water

Water used in mortar mixes should be pure enough to be drinkable. It should be clean and free of salts, acids, alkalis or large amounts of organic material.

Additives

In new construction, additives are sometimes used to modify chemical reactions in mortar in order to increase durability or improve working qualities. Modern chemical additives are, as a rule, unnecessary in rehabilitation and restoration and may, in fact, have a detrimental effect on old masonry walls.

Colors

It is best to achieve a color match through the careful selection of sand and binding material; this will produce the most consistent and permanent results. However, if it is not possible to obtain a proper color match, it may be necessary to use a mortar pigment. In fact, in the late 19th century some mortars were colored with pigments to match or contrast with the masonry units; red, brown and black pigments were commonly used.

If colors are required, chemically pure synthetic oxide pigments, which are alkali proof and sun fast, should be specified to prevent bleaching and fading. They should not exceed 10 percent of the volume of the binder (and in the case of carbon black, no more than 3 percent by weight). Organic dyes should not be used because they fade.

Antifreeze compounds

These additives (for example, calcium chlorides) tend to be detrimental to high-lime mortars because they introduce salts, which may later cause efflorescence (page 136), stimulate corrosion of metals in walls and even break down mortar.

Bonding agents

Although chemical, or bonding, agents can improve the bond with old mortar and adjacent surfaces, they are no substitute for proper joint preparation. They unnecessarily increase the strength of the mortar while making it less porous and more brittle; they can affect color (giving a bluish or greenish tinge), reduce breathability and, by dramatically reducing the ability of the mortar to weather, prevent the aggregate from naturally producing a proper color match to old mortar. In addition, it is difficult to clean off mortar mixed with these agents and smeared on the masonry surface.

Air-entraining agents

In northern climates, agents to increase the amount of air in cement mortars are sometimes used to improve their plasticity and resistance to freeze-thaw cycles, although these agents also may reduce bonding ability and compressive strength. A type 2A lime, which includes an air entrainer, can be used.

The Repointing Process

The difference between a good and a poor repointing job is not always obvious to the unpracticed eye. Merely brushing away the loose mortar and refilling the joint will produce a repointing job that may look good for several months, but within a few years the mortar will pop out of the joints. Good preparation of the joint takes a fair amount of work but is essential to getting a repointing job to last the 50 to 100 years that it should. It is during preparation for repointing that the masonry runs the greatest risk of permanent damage; cleaning out the joint should be done only by experienced workers using hand tools under the close supervision of an experienced mason.

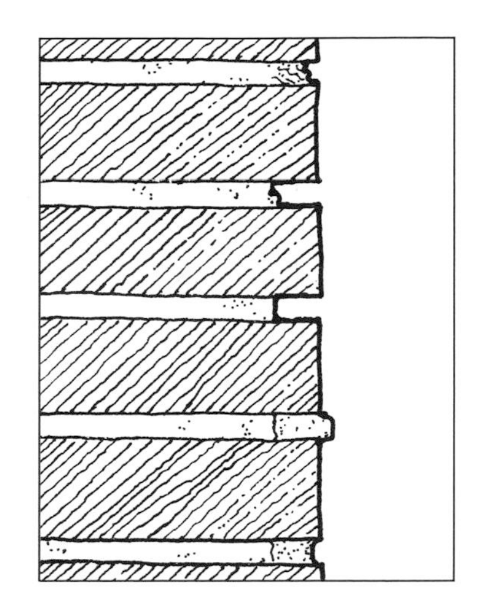

Joint preparation, top to bottom: Deteriorated mortar, mortar removed, joint properly raked, new mortar, tooled joint.

Preparing the Joint

All loose, crumbling, powdery, excessively soft, badly stained or cracked mortar should be raked (cut out) to a uniform minimum depth and the full width of the joint, preferably using hand rather than power tools.

Raking

To ensure an adequate bond, the joint should be raked to a depth equal to between 2 and 2½ times the width of the vertical joint (usually ½ to ¾ inch deep with brick and 1 to 2 inches with wider stone joints). Proper depth ensures that there will be enough surface contact between the mortar and masonry so that surface adhesion and friction will create a good bond without the use of special bonding agents. Any loose and deteriorated mortar beyond this minimum depth should also be taken out. Mortar should be removed cleanly from the masonry, leaving square corners and a flat surface at the back of the cut.

Before filling joints, any bricks or stones that are loose should be reset. Any pieces of brick that chipped off while chiseling out the old mortar can be glued back with ceramic

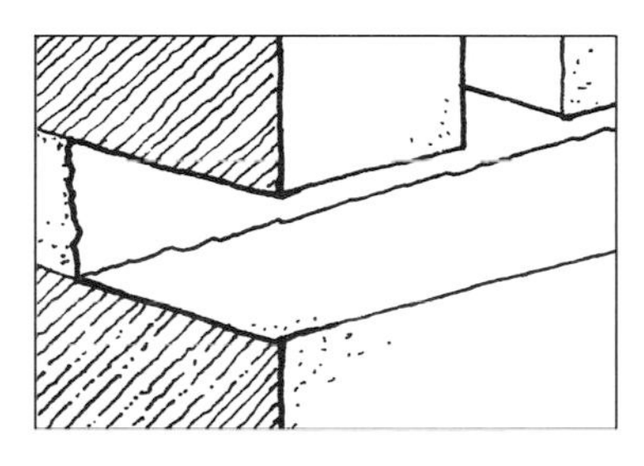

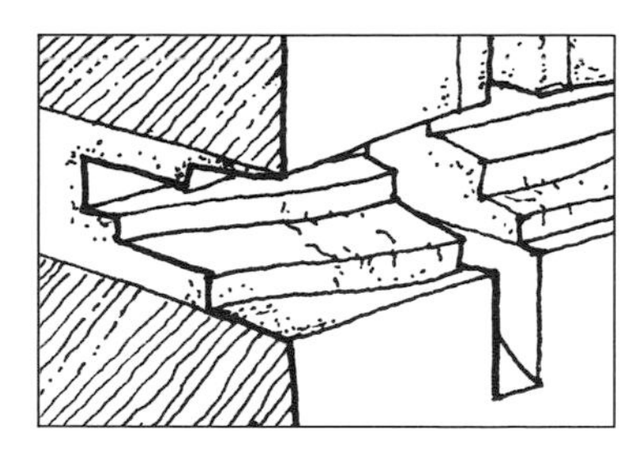

Top: Well-scraped joint showing an even face and depth of approximately 2½ times the joint's height.

Above: Joint scraped with a radial saw.

glue; stone can be reattached with epoxy. The joints should be finally cleaned out by gently flushing with water to remove all loose particles and dust. At the time of filling, the joints should be damp to prevent the too-rapid absorption of water from the new mortar, but no standing water should be present.

Top: Handraking, the best and safest way to remove old mortar. (Dean Korpan)

Above: Use of pneumatic tools, acceptable in certain situations when the right tools are used by experienced workers. They should be used rarely with vertical joints such as this. (Philip Marshall)

Hand versus power tools

The best way to remove old mortar is by hand using a small-headed chisel, no wider than half the width of the joint. Although handwork is more time consuming than using power tools, it presents far less risk of permanently damaging the brick or stone. If mortar can be removed only with power tools, it probably should not be removed in the first place.

For the most part, power tools such as circular saws with carbide blades or pneumatic impact hammers almost always damage the edges of the masonry units and overcut the ends of joints (especially the vertical joints in a brick wall). Damage to the brick or stone not only affects its visual character but can also lead to accelerated weather damage. Power tools may appear to do an acceptable job when the contractor does a demonstration, but when construction is under way and the day wears on, workers using power tools tire and the masonry inevitably suffers.

Power tools, if they are used to remove masonry joints, should be used only under the most controlled circumstances. Where joints are uniform and wide, it may be possible to begin the removal process using power tools, if the work is done by experienced workers under close supervision. For example, a small power grinder with a 4-inch blade can be used to cut the middle third of continuous horizontal joints; these joints should be finished and vertical joints done entirely with a chisel.

In certain extraordinary circumstances, such as removing portland cement from very narrow joints, very low-power, high-speed power tools in the hands of a skilled worker early in the day might not be as risky as using a hammer and chisel, but only if the mortar really must be removed.

Specifying that the mason or contractor replace all bricks or stones damaged during mortar removal with an exact match is one way to encourage that adequate care is taken to avoid damage (but it could have the unintended effect of encouraging carelessness). Power tools should not be risked on buildings of great significance.

No single mortar mix can be used for all repointing projects. The mix will depend on the circumstances, particularly the type of masonry and its exposure. Historical research on a building may turn up the original mortar recipe, which is usually a good guide in selecting a mix. Although a century-old mortar eventually deteriorated, this does not mean that anything was wrong with it; mortars have a limited life span.

The table accompanying this section gives an indication of appropriate mixes, based on the durability of the masonry material and the use and exposure of the wall. A pure lime-based mortar is quite slow to cure; while it initially sets in three days, it takes months to develop its strength fully. With soft historic brick and limited exposure, a little white portland cement can be added to speed up setting and improve durability. With hard brick and average-strength stones, the proportion of cement in the binder may be increased from 20 to 40 percent. With highly exposed granite, the cement might make up as much as 50 percent or more.

In specifying a mix of materials, proportions should be given in standard volumes, that is, 4 parts of lime to 12 parts of sand, rather than 2 bags of lime to 6 cubic feet of sand. Generally, the ratio is 1 part of binder (lime and cement) to between 2¼ and 3 parts of sand.

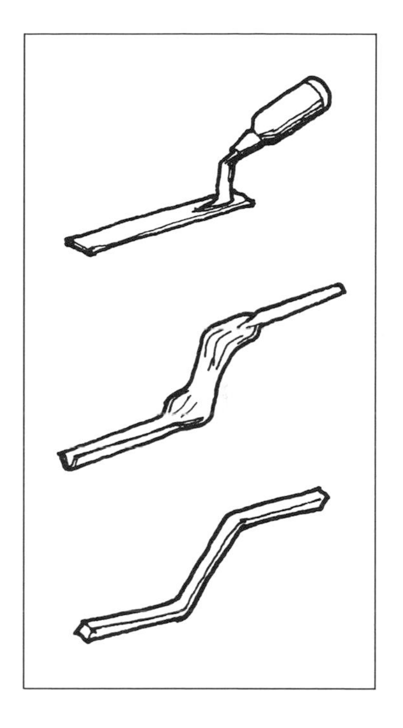

Repointing tools, from top to bottom: Repointing trowel, convex jointer or round iron, V-shaped jointer.

Matching the original mortar

Visual analysis of unweathered original mortar is usually sufficient to match the new mortar. The exact physical and chemical properties of the original mortar are not of major significance as long as the new mortar

- Matches the original mortar in color, texture and detailing
- Is softer (measured in compressive strength) than the brick or stone
- Is as soft (measured in compressive strength) as the original mortar

Even if an original mortar recipe is available, it may produce a different-looking mortar using the ingredients available today. Knowing the formula of the existing mortar, however, is a useful aid in developing an appropriate mortar mix for many rehabilitation and most restoration projects and is especially

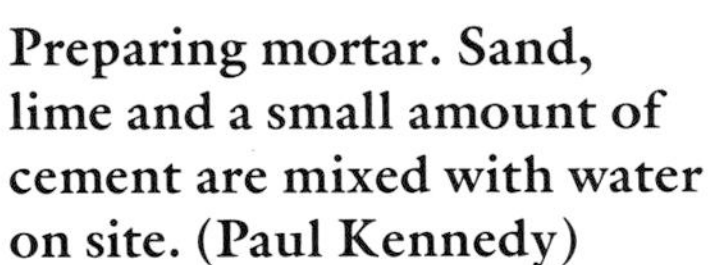

Preparing mortar. Sand, lime and a small amount of cement are mixed with water on site. (Paul Kennedy)

useful with buildings of special historical significance. Of the many methods for carrying out good chemical and physical analyses of mortars, a simple one that any mason or contractor can follow is outlined in the accompanying box.

To match the old and new mortars, select a broken-off sample of old mortar, snap it in two to expose its interior and compare it directly with a cured test sample of the new mix. Samples from several mixes can be made up to select the closest match. Setting can be speeded up in an oven. Alternatively, the old mortar sample can be wet down and compared with the wet mixed mortar: if they match, they should both dry to the same color. If the new sample matches the unweathered surface of the old mortar sample, it should eventually weather to match the mortar on the building exterior.

Mixing the mortar

Mortar should be mixed carefully to avoid lumps and uneven color and to ensure uniform strength and texture. Dry ingredients should be mixed first before adding any water. Half the water should be added, followed by mixing for approximately five minutes. The remaining water should then be added in small portions until the desired consistency is reached. The proper consistency uses the minimum amount of water to allow the mortar to stick to a trowel held upside-down. More water makes mortar easier to work, but it shrinks more, smears more easily and is not as strong. The total volume of water necessary may vary from batch to batch, depending on weather conditions. It is best not to work at freezing temperatures or, if this is unavoidable, to warm the sand and water and protect the completed work from freezing.

Mortar should be used before it begins to harden, generally between one and two hours of final mixing. The mortar should not be mixed in too large a quantity. Retempering, or adding more water after the initial mix is prepared, should not be done. A mix of lime and sand (no cement) can be made up in advance and stored in airtight containers; cement is added only when the mortar is used.

Basic Analysis of Historic Mortar

1. With a chisel, remove three or four unweathered samples of the mortar to be matched from several locations on the building. Because the masonry may have been repointed several times, it is important to remove several samples to obtain a mean mortar sample of the different mortars that have been used, avoiding obviously recent samples. Set the largest sample aside to be used later for comparison with the repointing mortar.

2. Break apart the remaining samples, powdering them with a wooden mallet until the mortar is separated into its constituent parts. There should be a good handful of the material.

3. To establish what the binder is, stir part of the sample into diluted hydrochloric acid. If there is a vigorous chemical reaction (bubbling) and most of the binder disappears, leaving clean aggregate, then the binder was lime. Cement will leave a murky liquid and will dissolve very slowly over several days.

4. To establish what the aggregate is, some must be isolated. Take the aggregate left in the previous step and rinse with water and dry. Alternatively, take more of the ground-up sample and carefully blow away the powdery lime or cement binding material; this will not work if the binder is too strongly adhered to the aggregate. Examine the aggregate with a low-power magnifying glass. Note and record the range of color as well as the varying sizes of the individual grains of sand or shell as well as the presence of other materials.

FILLING THE JOINT

The area to be repointed should be damp, but not wet, to slow down the absorption of water from the new mortar before it is properly set; otherwise, the mortar will not cure and adhere properly and thus will be weaker. Freestanding water or excessive dampness will delay the curing or cause excess shrinkage.

Wetting the area to be repointed and filling the joint. (Paul Kennedy)

Layering mortar

Ideally, the joint should be filled in successive layers, allowing each layer to harden before adding a next. Layering minimizes overall shrinkage, which can reduce the joint's watertightness.

Deeper joint areas — more than 1 inch — should be filled first, compacting the new mortar in several layers until the back of the joint is flat. Then, a ⅓-inch layer of mortar (a few feet long) is applied to the back of the joint, packing it well into the back corners. Several ⅓-inch layers will be needed to fill the joint flush with the surface of the masonry. Each layer of mortar should be allowed to reach thumbprint hardness before the next one is applied. If deep pointing is to be carried out in one operation without layering, the mortar should be stiff (not too wet) and well compacted.

To fill very narrow joints without smearing mortar on the masonry, the mortar can be inserted between two strips of waxed paper that are placed in the joint or masking tape can be used to protect the brick or stone.

Finishing the joint

Sometimes, masons finish joints simply by using a trowel to smooth out the mortar. This type of finishing, however, is not adequate and can actually make the surface more porous by creating a rough texture. To give mortar a smooth, denser outer layer, the joint must be tooled. Even if an untooled joint is being matched, it is generally best to tool and then let the joint weather or treat the surface so it matches.

Tooling

Tooling is the process of smoothing the joint with a finishing tool (slicker) narrow enough to be placed inside the joint. The slicker is pulled over the surface of the mortar to com-

Tooling the joint and brushing off excess mortar. (Paul Kennedy)

press it. By bringing the binder to the surface, compressing creates a slick film, often a different color from the rest of the joint. This film, which may initially inhibit water absorption and evaporation, can be allowed to wear off (usually within a year) or it can be carefully removed after pointing.

Proper timing of tooling is essential. If mortar is tooled when it is too soft, the color will be lighter than expected and hairline cracks may occur; tooling when too hard may cause tool burning (dark streaks) and prevent good closure of the mortar against the brick.

Ending the work

When stopping for the day, repointing should end at joints in the building, for instance, vertical elements such as pilasters or the edges of an arch or horizontal elements such as window sills or a stringcourse. In hot weather, a light misting will help slow down the setting and prevent suction from adjacent masonry. Burlap or a tarpaulin can be used to keep off the sun or heavy rain for the first few days.

Using a special tool to make a decorative bead on a foundation joint. (Frank Genello)

Shaping the Joint

The shape of the joint plays an important part in its efficiency and durability. Pointing styles used on the masonry and methods of producing them should be examined and reproduced to match the original joint profile.

A mortar joint should shed water to avoid the accumulation and penetration of rainwater between the mortar and masonry elements. From this point of view, the best profile is the concave joint; the worst is a joint that projects from the ma-

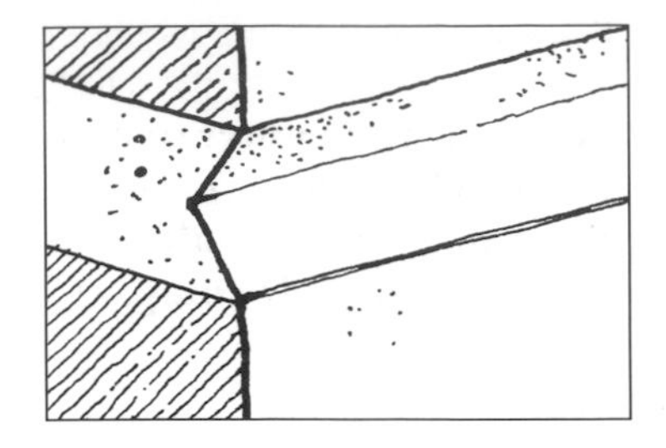

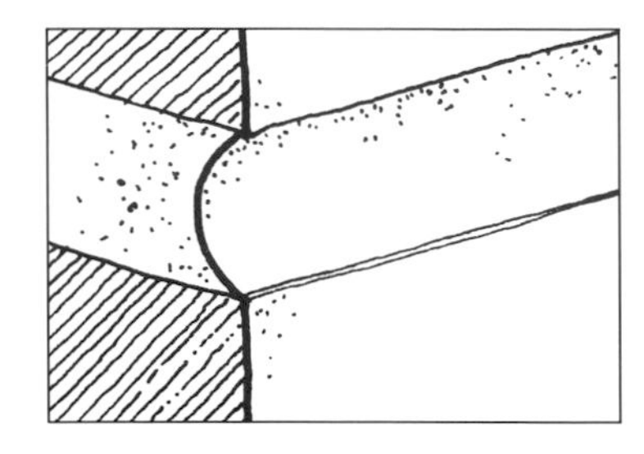

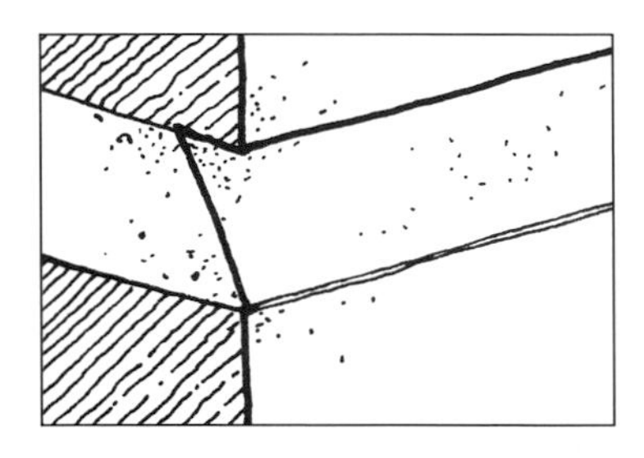

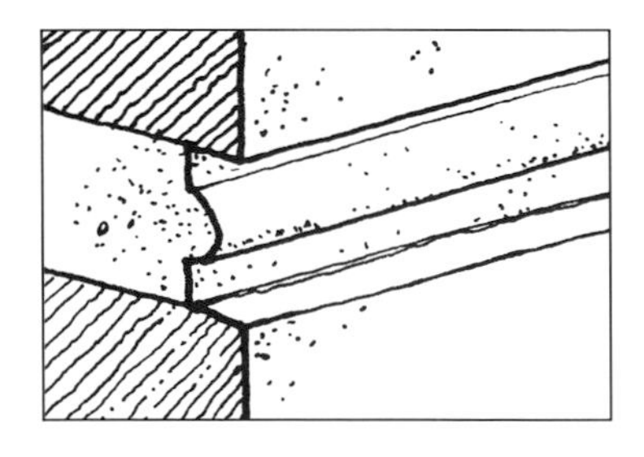

Some typical joint shapes, from top to bottom: V-shaped, concave, struck with drip, beaded.

sonry, exposing a wide surface of the mortar to weathering. If the original pointing had joint shapes that did not shed water properly or that had excessively thin edges that have broken off, it would be wise to select a joint shape that resembles the original, but is more technically sound.

The appropriate shape of the joint depends on the type of masonry. Because fieldstone and old brick have irregular edges, completely flush joints should be avoided, as they would leave delicate, thin, feathered edges that could easily break off and allow water to infiltrate; here, joints should be slightly recessed and concave. Flush joints may be appropriate only for thin joints with regularly cut stone or regular brick, but these, too, should be properly tooled, not merely scraped with a trowel.

Look at both horizontal and vertical joints to determine the order in which they were tooled and whether they were the same style. Some early 20th-century buildings have recessed horizontal joints while vertical joints were finished flush and pigmented to match the bricks, thus creating the illusion of horizontal bands. Pointing styles also often differed from one facade to another, front walls receiving greater attention to mortar detailing than side and rear walls.

Although the word "tuckpointing" is often used to refer to repointing in general, it originally defined a special kind of joint used to make wide, irregular joints (for example, in rough stonework) resemble thinner and more regular joints associated with costlier construction. The joint was first filled with a mortar colored to match the masonry units, then a narrow groove of about ¼ inch was made in the colored mortar joint and a strip of white lime mortar was "tucked" in.

Cleaning Up and Finishing

Carefully executed repointing should need little cleaning. Bits of mortar that fall off the trowel or are forced from joint edges by tooling are best removed with a stiff dry or lightly dampened brush after the mortar has initially set, but before it is hardened (often 1 to 2 hours, depending on the type of mortar). Hardened mortar can usually be removed with a wooden paddle or, if necessary, a chisel.

Smears on the wall should be cleaned up after a day or two, after the mortar has developed some resistance. This

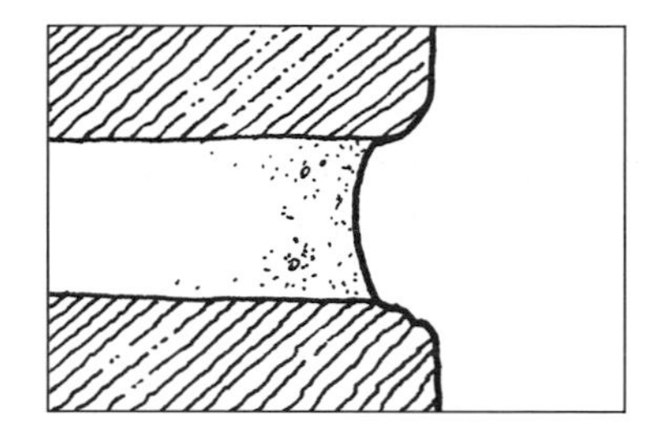

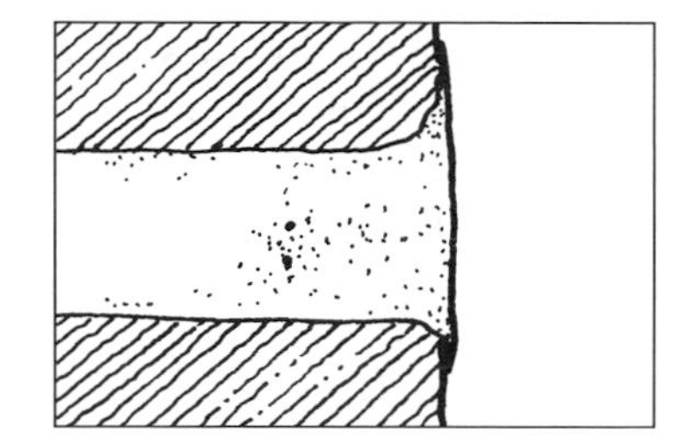

Packing the joint. Excess mortar leaves edges too thin, (left), allowing water to penetrate. This type of joint also will appear visually uneven and too wide compared to a well-packed joint (far left).

should be done with stiff, natural-bristle brushes and plain water. On glazed or polished surfaces, only soft cloths are appropriate. Because the binder of lime-rich mortars is acid soluble, acetic acid also may be applied with a small brush and flushed with water (see Cleaning Brick and Stone). Improper cleaning of a large area, such as with hydrochloric (muriatic) acid, can lead to mortar deterioration and discoloration as well as efflorescence.

If a full wash-down of the building is required to remove mortar bits (rare in rehabilitation), test panels should be used to evaluate the effects of different cleaning methods. New mortar joints are especially susceptible to damage, because they do not become fully cured for several months. The mortar should be completely hardened before a masonry cleaning project is undertaken; 30 days is usually sufficient, depending on weather and exposure.

In repointing work, new construction "bloom" (efflorescence, page 136) occasionally appears within a few weeks of repointing, although this is rare with lime-rich mortars. It should disappear through normal weathering.

Aging the Mortar

Even with the best efforts at matching the old mortar color, texture and materials, a difference will usually be visible, partly because the new mortar has been matched to the unweathered portions of the historic mortar. If the mortars have been properly matched, it is best to let the new one age naturally. No artificial aging technique should be used without careful evaluation and testing.

Various substances (ranging from solutions of potassium permanganate and carbon black to beer and manure) are commonly used to stain the new mortar. Staining is generally an unreliable and unstable technique; it may provide an initial

Repointed stone. The wall in the background was completely rebuilt and repointed using the original stone. Once the mortar has aged, its color will more closely resemble the original mortar in the foreground. (Paul Kennedy)

match but the old and new mortars may weather differently, leading to visual differences after a few seasons. Also, some mixtures used to stain the mortar may be harmful to the masonry (for instance, by introducing salts leading to efflorescence).

Tooling also may affect the look of the joint. The tooled patch may not match an adjacent weathered area where the lime or cement has eroded and the sand is visible on the surface. Also, a smoothly tooled joint may not be visually appropriate for a rough rubble wall.

Two options are available: to allow the joint to weather naturally, which should remove the film in a few months, or to remove the film by rubbing very gently with a damp, soft-bristle brush, such as a toothbrush, or with sacking. Techniques that seriously abrade the surface of the mortar should be avoided.

Scrub Coating

The terms "slurry coating" and "scrub coating" are used by contractors to describe new techniques that involve brushing a thinned, low-aggregate coat of mortar over the entire masonry surface; when dry, this coat is scrubbed off the brick or stone with a brush, presumably leaving a residue of mortar in the joint. Other methods — "mask and grout" or "tape and grout" techniques — call for taping the edges of the joints to protect the masonry and brushing the slurry into the joints with a brush. These techniques may seem appealing because they are quick and inexpensive compared to traditional repointing and do not require skilled craftsmanship. However, they should not be confused with or substituted for true repointing and are especially inappropriate for historic buildings.

Scrub coating may be of limited use in sealing hairline cracks in the mortar, particularly with very fine joints where repointing would be difficult. For the most part, these superficial cosmetic techniques do more harm than good. They tend to mask joint detailing or tooling, have a life expectancy of only a few years and may be extremely difficult to clean from the surface of the brick without leaving a residue, called veiling.

Guide to Mortar Mixes

This table gives an indication of the proportion of various mortar ingredients based on the role and location of the masonry as well as the strength of the stone or brick. The letters O, N and S correspond to the equivalent ASTM standards. Type M (very hard) mortars, which are not indicated here, are made with very little lime and are too hard for use with old masonry.

Use | **Strength of the Masonry** | **Mortar Mix**

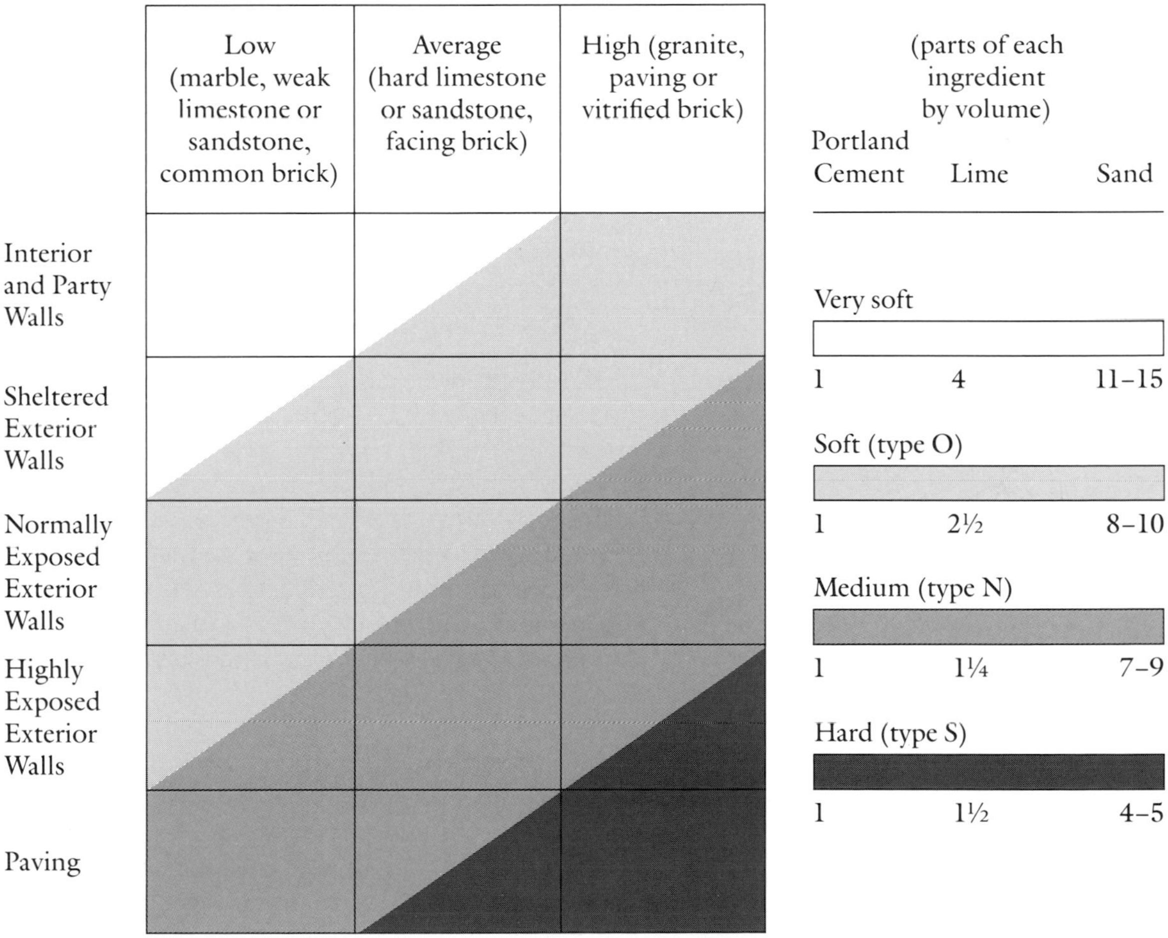

Use	Low (marble, weak limestone or sandstone, common brick)	Average (hard limestone or sandstone, facing brick)	High (granite, paving or vitrified brick)
Interior and Party Walls			
Sheltered Exterior Walls			
Normally Exposed Exterior Walls			
Highly Exposed Exterior Walls			
Paving			

(parts of each ingredient by volume)

	Portland Cement	Lime	Sand
Very soft	1	4	11–15
Soft (type O)	1	2½	8–10
Medium (type N)	1	1¼	7–9
Hard (type S)	1	1½	4–5

Repairing Deteriorated Surfaces and Building Parts

The types of problems resulting from masonry deterioration run the gamut from fairly insignificant examples such as erosion of the stone or brick surface to serious conditions such as cracked or bulging walls.

The first step in treating the deterioration of masonry is to arrest the decay by finding and eliminating the cause of the problem. Then, it may be desirable to treat the masonry so that it will be less susceptible to further damage. The final step in some cases is to restore the masonry's original appearance.

If the degree of deterioration is minimal, and the problem does not threaten the structural integrity of the building or detract too much from its architectural character, it is preferable to leave the masonry alone. Old buildings cannot and should not be expected to look perfect. If repairs are to be carried out, as much of the original material as possible should be kept.

Unfortunately, technology has not advanced to the point where there is a treatment for every problem or a standard formula for all situations, even when a solution does exist. Because buildings are complex and masonry materials vary, apparently similar problems affecting seemingly similar brick and stone can have different causes and effects. Therefore, treatment must be carried out on a case-by-case basis and only after testing.

Some treatments have been used successfully for a long time, but many new treatments have not; their application may do more harm than good. So approach new "miracle" solutions very cautiously.

Stone suffering from most forms of deterioration can be repaired using the techniques described in this chapter, although in some situations badly damaged stone must be replaced (page 154). Unfortunately, individual bricks generally cannot be repaired and if badly damaged must be replaced or covered.

Water damage to sandstone. Poor drainage led to this surface erosion. (Michael F. Lynch)

Opposite: Scaffolding around a Victorian home under rehabilitation in Georgetown, Washington, D.C. (Paul Kennedy)

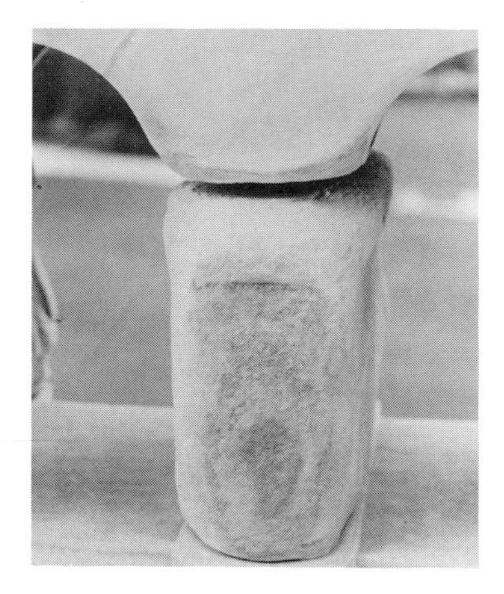

Weathering of marble. (John H. Myers)

Virtually all of the techniques discussed in this chapter should be carried out by masons or contractors experienced in masonry restoration and under the supervision of a qualified expert.

How Brick and Stone Decay

Brick and stone decay for many complex and often interrelated reasons. The main causes of deterioration are a variety of external forces — physical (stress caused by structural movement, frost action, crystallization of salts, thermal change), chemical (acid rain) and biological (plants, fungi). The effects of these forces will be greater if defects exist in the masonry, such as high porosity or the presence of weak strata in sedimentary rocks. Construction errors such as incorrect bedding of sedimentary rocks, use of iron anchors that later rust, improper use of soft, low-fired brick on an exterior wall, and poor renovation practices such as sandblasting and the use of too hard a mortar in repointing also contribute to deterioration.

With the help of the following photographs and descriptions, it may be possible to identify what kind of deterioration is affecting the brick or stone. While the definitions of some terms sound similar, such as those for exfoliation, peeling and spalling, the phenomena usually are brought about for different reasons.

Erosion of sandstone steps. The erosion follows the bedding planes of the sandstone. (Anne E. Grimmer)

Pitting of limestone. (John H. Myers)

WEATHERING OR EROSION

Weathering, also called erosion, is the general term for natural disintegration of the surface, edges, corners and carved details of masonry. It is caused for the most part by conditions such as wind, windblown particles and rain.

Weathering is particularly pronounced on sharp corners and highly carved or projecting architectural details, where it results in granular and rounded surfaces. With sandstone, the grains of sand stand above the surface and can be rubbed off when touched, indicating that the cementing material has been lost. Acid rain can increase natural weathering rates, resulting in noticeable softening or loss of masonry details, particularly with acid-soluble, carbonate stone. Weathering can show a uniform pattern across the surface or can produce varied patterns.

Honeycomb, or alveolar, weathering is a type of erosion common to sandstone and limestone surfaces in which cavities (alveoles) are created in a honeycomb pattern. It occurs primarily in arid climates where strong winds encourage the evaporation of salts directly below masonry surfaces, but it can also be found in more humid areas.

Pitting, the development or existence of small cavities in a nonhomogeneous masonry surface, results from differing rates of erosion of individual particles in masonry. It may be caused by natural weathering or harsh or abrasive cleaning methods.

Salt fretting, sometimes called salt erosion, is a pattern of erosion or etching caused by salt, usually from the salting

Salt fretting of sandstone. (Michael F. Lynch)

Sugaring of marble. (Anne E. Grimmer)

of icy sidewalks. Unless the use of de-icing salts is discontinued, this condition may eventually result in spalling and exfoliation of the stone surface.

Sanding, or sugaring, is a granular, sometimes powdery condition characteristic of fine-grained marbles. It is associated with gradual surface disintegration.

Weathering is particularly difficult to arrest. Minor damage is probably best left alone; with serious, progressive damage to areas of significant detail, the use of a consolidant (page 152) might be considered.

Surface Crust

Many types of decay are related to the presence of a hard, thin surface crust or skin. This crust may be the result of one or more of the following factors:

- Architectural finishing treatment of stone, during which the surface is worked using techniques such as bush hammering. These techniques make the outer part of the stone more dense.
- Movement of moisture toward the surface of stone, which leaches cementing materials, salts or other substances from

Right: Surface crust on sandstone resulting from dissolution of the stone. (Erhard M. Winkler)

Far right: Protective coating peeling from brick. Here, the coating does not appear to be removing the masonry surface. (Ward Bucher) within the stone and redeposits them on the surface. Some of these crusts, such as calcitic ones in limestone, can provide a protective surface to the stone.

- Chemical reaction between the stone and airborne pollutants, particularly calcareous stones. Calcitic binders in limestones, marbles and sandstone react with acid rain to create a crust; such crusts may be more fragile than the rest of the stone and may indicate that the stone could soon disintegrate.
- Previously applied and unsuccessful protection treatment. If not well adhered to the masonry, only the coatings (such as paint or stucco) may flake or peel; but if a coating is well adhered, the masonry surface itself may separate.

Note that the formation of surface crusts may take place both across and parallel to the bedding planes of sedimentary stones, a clue to differentiating this phenomenon from exfoliation, in which layers are separated parallel to the bedding plane.

Problems arise in surface crusts because the thin, surface layer may behave differently from the rest of the material when subject to wetting and drying and heating and cooling or to the effects of freezing moisture. As a result the crust can blister, flake and peel and eventually separate from the surface, perhaps pulling some underlying material with it.

Blistering occurs when the surface swells and ruptures. It is rarely associated with natural crusts and is usually the result of applied treatments. Because blistering can be triggered by de-icing salts and ground moisture, it generally is found on a surface close to the ground. Blistering may remain a relatively constant condition scattered over the masonry surface, or it may be the first symptom of a problem that may worsen.

Flaking is the detachment of small, flat, thin pieces of the outer layers of brick or stone from a larger piece of building stone. It may be an early stage of exfoliation.

Peeling is the loss of larger areas of the stone's surface crust.

Most problems related to surface crusts are difficult to treat, and the problem areas usually can be left alone unless the damage is so serious that it threatens the masonry unit. It is sometimes possible to alleviate crust problems caused by the previous application of surface treatments, such as waxes, by removing these materials.

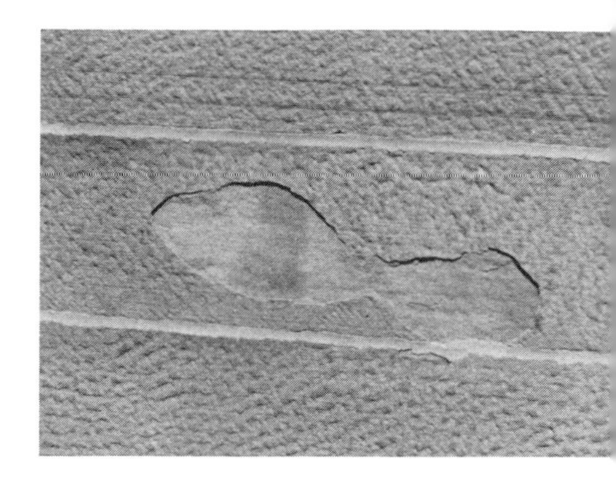

Top: Blistering of sandstone. (Anne E. Grimmer)

Center and above: Flaking of granite and peeling of granite. (Baird M. Smith)

Exfoliation or Delamination

Exfoliation, also referred to as delamination, is the splitting apart of the outer surface of natural stone into thin layers (laminae) that peel off.

Because of their layered composition, this may be a natural condition of sedimentary stones such as sandstone, where delamination takes place along the natural bedding planes of the stone; the presence of clay or mica-rich layers may accelerate the process. Sedimentary stones tend to delaminate more quickly when they are improperly laid.

Blind exfoliation is a form of partial exfoliation indicating that the surface layer is only loosely attached and has not yet separated. When tapped with a rubber mallet, stones with this condition will sound hollow.

To treat exfoliation, the first step — as with all masonry repair — is to remove or secure unsound material that presents a danger of falling off. Then, the problem should be diagnosed and the cause treated; for example, if excessive moisture is the cause, the source must be controlled (see Finding and Treating Moisture Problems).

Finally, one can deal with the appearance. If damage is not deep, the surface should be left as is. With significant deterioration, it may be possible to patch using mechanical or plastic repair techniques (pages 145 and 148).

Top: Delamination of sandstone along the bedding planes. (U.S. Army Corps of Engineers)

Above: Exfoliation of sandstone. (Anne E. Grimmer)

Efflorescence and Subflorescence

Efflorescence is a whitish haze of soluble salts left as deposits on masonry by the evaporation of water. The soluble salts may come from the wall itself (for example, sulfates inherent in the raw materials of some bricks or carbonates resulting indirectly from the use of portland cement) or may be drawn into the masonry from salty ground water. In addition, airborne or water-deposited pollutants in the atmosphere may cause sulfates to appear on the surface of the masonry. Efflorescence also may be a combined salt residue left on the masonry surface by excessive washing or chemical cleaning, often from too strong a chemical solution or inadequate rinsing.

Efflorescence itself may be more unsightly than harmful, but its presence on an old building serves as a warning that

Above left: Efflorescence and spalling on a brick wall. (Anne E. Grimmer)

Above: Spalling of brick. (Susan Dynes)

water has found a point of entry into the structure and that more serious damage may follow. Efflorescence may also indicate potentially damaging subflorescence; this is an accumulation of soluble salts beneath the masonry surface, which may exaggerate the effects of other conditions such as exfoliation and spalling, eventually leading to deep deterioration of the masonry.

After the source of excessive moisture has been identified and eliminated, efflorescence can be removed with dry brushing or water washing (page 97). Efflorescence and subflorescence can be removed with a poultice (page 105). Both conditions are discussed in more detail in Finding and Treating Moisture Problems.

Spalling

Spalling is the breaking away of the outer portion of a masonry unit in a localized area as a result of the pressure of salts and of cyclical freezing and thawing of moisture trapped under the surface. The term is sometimes used loosely to refer to any coming away of the surface, including exfoliation. Too hard a mortar also leads to spalling. Pressure from movement in the wall is deflected, rather than absorbed, by the mortar joints and placed instead on the masonry units, eventually causing them to spall. In addition, spalling may result from deterioration of the internal anchoring system that holds the units to the structure; water infiltration causes metal anchors or reinforcements to rust, creating increased internal pressure.

Chipping, the separation of small or large fragments of masonry from the masonry unit, often occurs at corners of

Top: Severe spalling of brick. (© Richard Pieper)

Above: Chipping of granite. (Anne E. Grimmer)

mortar joints. It may be caused by later alterations or repairs, such as use of too hard a mortar, or by accident or vandalism.

After the cause of the spalling has been identified and, if possible, halted, one can deal with the surface appearance. If damage is not deep, the surface should be left as is. With significant deterioration of stone, it may be possible to use mechanical or plastic repair techniques, such as a dutchman or composite patching (page 148). Spalled brick should be replaced. In a few cases water-repellent coatings or consolidants may be used.

Cracking

Cracking, identifiable as narrow fissures (1/16 inch wide or more) in a block of masonry, may result from a variety of conditions, such as freezing and thawing, structural settlement of a building or the presence of too hard a repointing mortar.

Small cracks within a single block of masonry may not be serious, but longer and wider cracks extending over a larger area may be indicative of structural problems and should be monitored.

Once the cause has been dealt with and the situation is stable, large cracks in a stone can be repaired by injecting epoxy (page 147); minor ones can be filled with grout.

Surface Treatments

Surface treatments traditionally have been applied to masonry to solve deterioration problems and particularly to try to keep out moisture. But sealing the surface of masonry is more likely to trap moisture inside the wall than keep it out.

Moisture problems in a wall rarely come from penetration of rain through the surface of the masonry material itself, unless it is cracked; such problems are far more likely to result from open joints, ineffective roof flashing, poor seals around window and door frames, missing or damaged gutters and downspouts, rising damp or condensation. In addition, strong forces push moisture outward, particularly evaporation from the surface.

Cracking. Crack in a dutchman repair (left, © Richard Pieper) and in limestone (far left, John H. Myers).

Because trapping moisture usually creates a more serious problem than penetration, the recent widespread use of waterproof sealers, which prevent moisture from entering or leaving the masonry, and water-repellent coatings, which only repel water but continue to allow moisture within the masonry to evaporate, does not appear warranted and could lead to even more serious damage. In addition, these coatings are difficult and costly to remove.

On the other hand, many paints may not be as harmful because early stages of failure can be detected more easily and paint can be more readily removed should it cause problems. Use of stucco, tiles, shingles and other wall coverings may be considered where there is historical precedent for the application.

In general, the use of masonry surface coatings should be avoided except in special circumstances where a badly spalled building risks water penetration; brick walls have been sandblasted; a masonry surface is subject to heavy, wind-driven rain; or parapet copings have deteriorated (as a temporary measure until brick can be replaced). Even in these cases, coatings should be used only after all other remedies have been exhausted and after consultation with an independent expert and extensive testing. It is, of course, appropriate to use paint if, historically, the surface has been painted.

Waterproof Coatings

Waterproof coatings, or sealers, make a masonry surface impermeable to water by sealing it from both water and water vapor. They may be clear (acrylics, epoxies, polyurethanes) or opaque (bituminous coatings, tar, pitch, hydraulic clay, and asphaltic, aluminized and other waterproof paints).

They may be used effectively on the dry exteriors of foundations and basement walls below grade in conjunction with drainage measures and also in association with damp-proof courses to prevent rising damp (page 190). Trapping moisture within the wall is unlikely in this case as no evaporation occurs below grade.

However, because water inevitably enters a wall above grade, the presence of an impermeable coating — even one just above grade — may still lead to serious damage. Trapped water will freeze and eventually escape to the interior, damaging finishes, or build up pressure behind the coating, causing the surface to spall.

Waterproof coating that has trapped water. Water pressure behind the coating forced the coating to peel, bringing with it part of the stone's surface. (National Park Service)

The materials used for waterproof coatings are generally solids applied in solution. As the solvent evaporates, the material is deposited on the surface of the masonry, creating a thin film only a few millimeters thick. Even minor differences in the thermal and moisture properties between the film and the masonry set up stresses that can lead to separation of the film.

Waterproof coatings sometimes become patchy-looking immediately after application or after a few years. Unless the coating is regularly renewed (every five years), patchiness will result inevitably from the wearing away of the coating from parts of the surface, causing uneven wetting patterns.

Water-repellent Coatings

Water-repellent coatings create a surface that is resistant to and repels water but is not impervious to water. Such coatings allow water vapor to enter and leave through the pores of the masonry. Manufacturers often call them "breathing" coatings.

Traditionally used repellent coatings such as waxes and fats have been replaced by silicones and metallic soaps (for example, aluminum stearate). The new water-repellent coatings are colorless and transparent, although some can give masonry a hazy look by changing its reflective property.

A problem with many of these so-called breathable treatments is that they tend to block up with reapplication, actually becoming impermeable even to the movement of vapor. Many water-repellent coatings must be applied on a completely dry wall. This can present a problem as they are most likely needed with a chronically damp wall.

Even if they did not harm the masonry, water-repellent coatings may not be cost effective. They are relatively expensive and must be renewed. Although they are often accompanied by impressive guarantees, the fine print usually will show that the liability of the manufacturer or contractor is limited — perhaps only to the cost of materials and for a few years.

Cleaning contractors often suggest applying water-repellent coatings routinely after cleaning, "to prevent the wall from getting wet and dirty." This automatic application is unnecessary and inappropriate. The few years that a protective treatment may last are insignificant compared to the many years a cleaned brick or stone wall will last before needing another cleaning.

Repellent coatings should be considered only as a last resort and only for areas where there is evidence that the masonry is actively deteriorating and all other means to prevent further moisture damage to a masonry surface have failed — for example, with sandblasted or badly spalling brick or sandstone. Depending on the cause and severity, the application of a water-repellent coating to a limited area may slow down the rate of deterioration, although it will not prevent further spalling and will be, at best, a temporary solution.

Water-repellent coating, evident from its blotchy appearance. (Walter Smalling, Jr.)

PAINT

Paint has traditionally been applied for decorative purposes or as a protective coating for poor-quality or porous brick. It creates many of the same problems as other coatings and should also be added only as a last resort to lessen moisture-penetration problems, although its reversibility sometimes makes it preferable to water-repellent coatings. If a surface is actively eroding, paint will not adhere well.

Only a paint that allows vapor transmission should be used. Whitewashes (lime washes) have been used traditionally with certain success, although some latex paints formulated especially for exterior masonry walls are now preferred; some of these are made with various types of aggregate or thickening agents to smooth rough or uneven masonry walls. Remember that the use of many different types of paint will make later removal more difficult. Any paint that creates an impermeable film (epoxies and most alkyds) must be avoided.

PARGING AND STUCCO

In the early days of North American settlement, some masonry walls, particularly those of rough-cut stone, were often covered with mortar or stucco either to protect badly deteriorated walls or to decorate them.

The word "parging" (rendering) usually refers to the application of a thin coat of mortar directly on a masonry wall, particularly as a damp-proof measure for rough masonry, foundations and basement walls; today, cement mortar is used and often contains damp-proof ingredients. "Stucco" generally refers to a thicker coat built up in several layers either over masonry or nonmasonry walls, most often on a mesh or furring strips; the term "plaster" is used in some areas for historic stucco coatings that do not contain any portland cement. The terms "stucco" and "parging" occasionally are used interchangeably.

Traditionally, stucco consisted of various mortarlike mixes of clay soil, lime, animal hair, straw, pigments and sand, all mixed with water. It was sometimes whitewashed or could be scored to resemble cut stone, making a rough stone or even brick building look as though it had been built of more expensive materials. Because traditional stucco was lime based, it could generally tolerate movements in the wall; moisture also could move through stucco and evaporate. Stucco made of portland cement is brittle and has much lower porosity; it easily develops fine cracks that then draw moisture into the wall through capillarity and traps it inside, leading to failure of the stucco finish.

Stucco is often used on the lower portion of a wall to treat moisture problems. Although it can be of limited use in reducing the back-splash of rain from the ground onto the wall, it also prevents rising damp (moisture) from evaporating,

Top: Stucco over brick, scored to resemble ashlar. (© Richard Pieper)

Above: Parging over brick. (National Park Service)

Below: Repairing stucco. Loose stucco is removed down to the scratch coat or the masonry itself; the area is undercut; and the stucco is applied in layers.

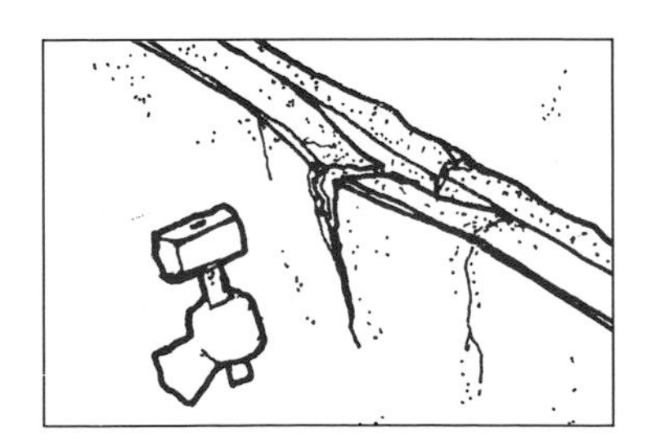

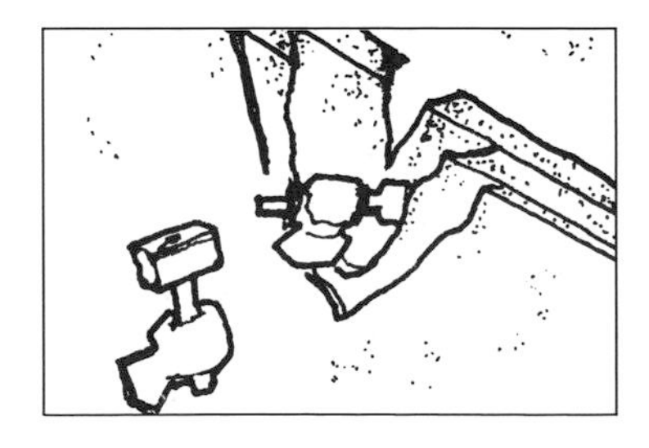

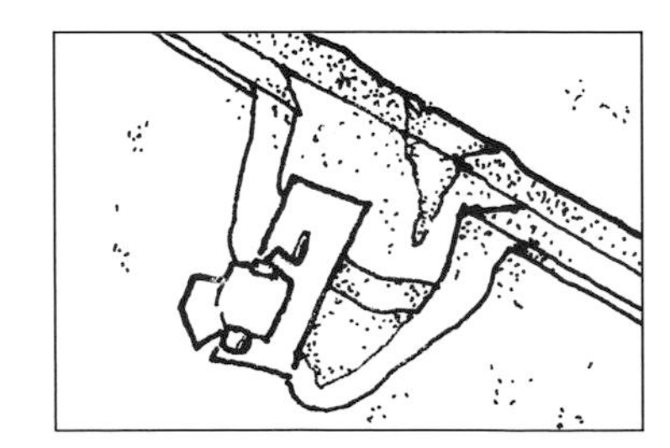

spreading the problem even more (page 178).

Stucco is generally applied in three coats, thus reducing cracking and making a smoother final coat possible. It is best laid on galvanized metal attached to furring strips to create an air space between the stucco and the original masonry wall to help prevent further moisture damage.

Repair

A stuccoed masonry wall that is badly deteriorating may tempt a homeowner to remove the stucco completely to expose the brick or stone. Remember, however, that the stucco was probably applied for a good reason; an unpleasant surprise may await underneath. It is particularly difficult to remove stucco from a brick wall without damaging the wall. Most stucco repairs require only patching, which should always be done if stucco was the original treatment for a historic building.

Loose areas either will be evident or may be detected by tapping the wall with a wooden mallet. The loose stucco will sound hollow. Loose or bowed-out areas are then removed down to a sound surface, the scratch coat (first layer) or the masonry itself. Because it is hard to disguise a patch, replace sections of a wall between logical break points. On the other hand, as much of the original material as possible should be kept, so the smallest possible area should be patched, especially on historic buildings. Cracks should be cleaned out, and all edges of sound mortar around the patches and along cracks should be undercut to allow the patch to grip.

Patches are laid in several layers directly on the masonry itself or on galvanized metal lath attached to the masonry. The patching material should look the same and have chemical and physical properties compatible with the original, for example, a soft lime and sand stucco should not be patched with portland cement. Mixes are similar to mortar mixes (page 115). A typical mix is 2 parts of lime to 1 part of portland cement and 9 parts of sand, with a little animal hair. Two base coats, 3/8 to 1/2 inch thick, are first applied: the scratch coat, followed the next day by the second or "brown" coat. Each coat is scratched to improve the bond with the next coat, and the wall should be wet down before applying each layer. A week later, the finish coat is applied — made up of a higher lime mix to prevent shrinking and cracking. This final 1/8- to 1/4-inch coat is

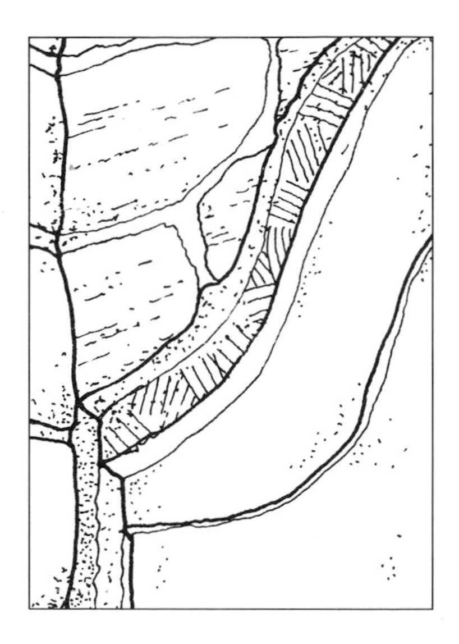

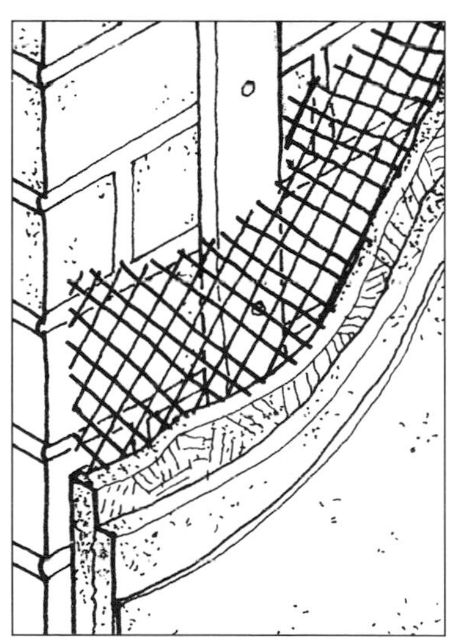

Top: Stucco applied in three layers directly on masonry. The first coat is scratched to give better adherence.

Above: Stucco on a galvanized metal lath nailed to wood furring strips, which create an air space to prevent humidity from damaging the wall.

Top: Protecting an exterior surface through the winter before completing stucco work in the spring. (Restoration Workshop)

Above: Metal siding covering the original brick surface of a row house. (Paul Kennedy)

given a smooth or textured finish to match the rest of the wall. Avoid too rapid drying by working on an overcast day or late in the afternoon or by misting with a hose.

To allow the surface to breathe, it is best not to paint or seal stucco. Tints should be used to match the original color.

New stucco

Applying a new stucco finish on a building that did not previously have one would not be acceptable on the main facades of a building, particularly one of historical importance. It might be acceptable on walls of no architectural significance (for instance, the side walls of a row house exposed because the adjacent building is destroyed) or as a temporary measure for a wall that is so badly deteriorated it must later be totally rebuilt.

Siding

In some northern areas, shingles, slates and wood siding were traditionally used to protect masonry walls, sometimes only on the most exposed wall of the building. Where there is no historical precedent, however, such treatments would be completely out of character with the masonry building and should not be added. Metal and vinyl siding are particularly inappropriate: they are aesthetically incompatible; they may trap moisture in the wall, worsening damage; and they are easily damaged by impact, especially near ground level, after which they must then be replaced because repair is virtually impossible.

Stone Repair

Bricks generally are not repaired because it is easier to replace damaged units. Stone, however, lends itself better to patching and other kinds of repair. Not all deterioration or damage can be or even needs to be repaired. The value of a building rests in its original materials, and it is best to conserve materials by stabilizing deterioration rather than undertaking the irreversible act of cutting out and replacing. Thus, if the damage caused by the wearing away of the surface from weathering or blistering is only slight — less than a few inches wide — it may be best to leave the masonry as it is.

If a few stones in a wall are partially missing, broken, cracked, chipped, badly weathered or otherwise deteriorated, they often can be replaced or patched. Most of the stone repair techniques described here serve essentially cosmetic purposes; particularly with buildings of notable significance, it may be better to conserve the somewhat weathered or damaged original materials. On the other hand, damage that threatens the structure, such as cracks that allow water to penetrate into the wall, must be repaired.

If a new piece of the same or a compatible material is mechanically attached with glue, epoxy or pins, it is called a mechanical repair. If the patch is a malleable material that hardens in place, it is called a plastic repair or composite patch.

Mechanical Repair

Fractured masonry often can be refastened or replacement pieces can be attached with mechanical repairs. These techniques may be appropriate for use on stone that has cracked, delaminated or exfoliated, as well as on pieces of masonry that have become detached from the wall.

All mechanical repair techniques are similar. They involve attaching a piece of masonry to a surface to replace damaged or missing material. The piece can be either the original material, a new piece of matching stone or a stonelike substitute such as artificial stone (a small piece of precast concrete made to resemble stone).

The remaining original masonry is first cleaned out and cut back, and holes are drilled or grooves chiselled to provide a key for the bonding material. The prepared surface is cleaned free of dust and stone chips using a stiff nonmetallic masonry brush. All stone and other surfaces adjacent to the area being repaired should be protected with rubber cement, which is removed after the repair has been completed. The mortar joints affected by the repair work should be raked out and repointed after the work is done.

Common bonding materials are cement grouts and epoxy. Epoxy resins are particularly good for reattaching small, carved details or other decorative elements and for making small dutchman repairs (page 147). Epoxy should not be used

Detachment of marble from a pediment, suitable for a mechanical repair. (Walter Smalling, Jr.)

Reattachment. This newly carved sandstone bird's head replaces one lost through vandalism. The head is reattached using stainless-steel pins and epoxy. (Pietra Dura, Inc.)

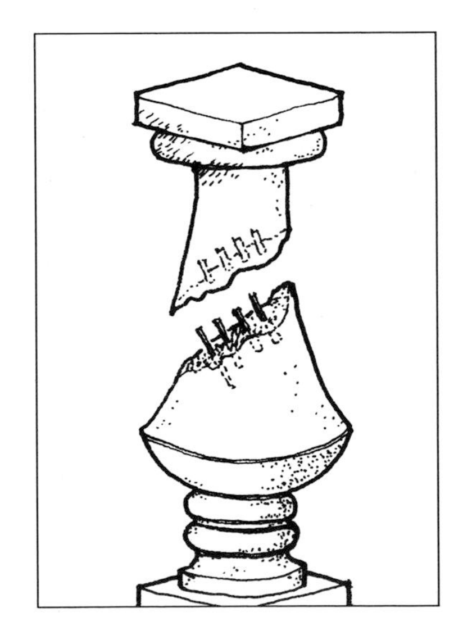

Typical reattachment using pins and epoxy.

for large pieces, as the impermeable layer would trap moisture. In these cases (for example, large dutchman repairs and veneering), anchors, such as metal clips, should be used to attach the new piece mechanically, or the piece may be set in a fine lime mortar grout, which would allow moisture to escape.

Epoxy resins used in masonry repair should have the consistency of thin paint, be rigid when cured and be insensitive to moisture and resistant to solvents, acids and alkalis, especially if the building is to be cleaned with these products afterward. The manufacturer or the architectural expert should be consulted on the best epoxy resin for the job. A hardening agent (B part) is added to the resin (A part) so that it may harden; it must then be used before it sets (between ½ and 1½ hours, depending on the type of resin) and must be kept at a relatively constant temperature and protected against the weather while it cures (between 1½ and 6 hours). Most epoxy resins are translucent, with a slight yellow tint. Adding dry, sifted stone dust to the resin and hardener mix helps match the resin to the rest of the masonry.

Reattachment

With this procedure, also called concealed repair, detached pieces of stone are held in place by adhesive and pins. If possible, the parts to be glued should be removed from the wall to facilitate working on the broken surfaces. All surfaces should be clean of dirt, stone dust and oil. For lintels and other projecting and bearing members, reinforcing pins of stainless steel, bronze or thermoplastic can be used. Traditionally four to six times its diameter in length, each pin should be scored or threaded to provide a gripping edge for the adhesive. Holes are drilled in both the base piece of masonry and the replacement piece. The depth of a hole is two to three times a pin's diameter and the width is ⅛ inch larger than the diameter. Epoxy is placed in the holes, the pins are set in the base piece, more epoxy is applied uniformly on one of the broken surfaces, and the two pieces are joined together. All surplus glue is cleaned immediately, and the joint is left to dry, protected from the weather.

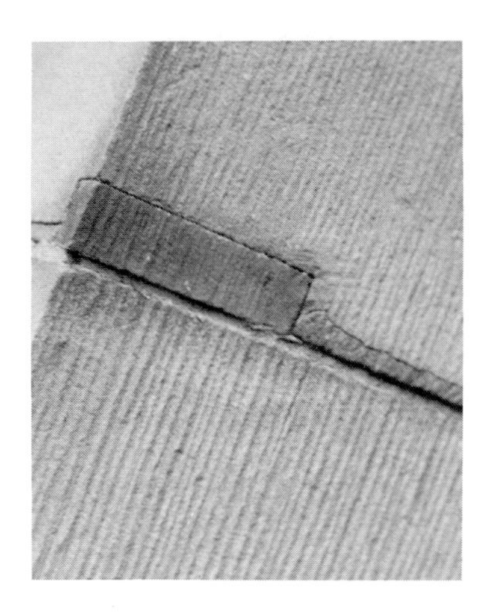

Dutchman repair. Ideally, the color of the patch stone should match the original stone. (Anne E. Grimmer)

Dutchman repair

A dutchman is the piecing-in of a small patch of natural stone or precast concrete imitation stone as a treatment for chipped or damaged stone. The piece is held in place with an adhesive such as epoxy and attached in a procedure similar to reattachment. The joint between new and old should be kept as narrow as possible to maintain the appearance of a continuous surface.

Crack repair

There are various sophisticated techniques for repairing cracks with epoxies. These are difficult to execute and should be attempted only by expert masons. Small cracks are covered with nonoily modeling clay to prevent leaks, then epoxy is injected with a syringe. For large cracks, a series of holes is made in the clay, epoxy is injected into the lowest hole until it comes out the next higher one, then the first hole is plugged and the epoxy injected into the second, working up to the top; after drying, the clay is removed and the surface of the crack filled with the same kind of mix used for plastic repair (page 148). Another technique is to first fill the surface of the crack with a plastic repair mix such as lime and stone dust, imbedded with small plastic tubes. Epoxy is then injected through the tubes deep into the crack; finally, the ends of the tubes are cut off. These techniques are similar to the epoxy injection treatment used to repair cracks in foundation walls.

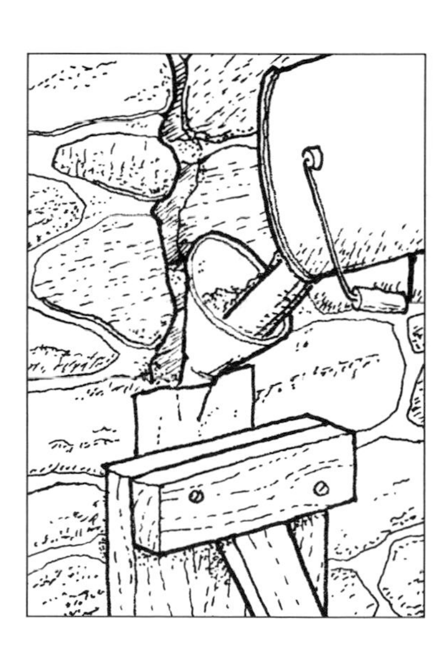

Crack repair with cement grout. Temporary formwork keeps the grout in place as it hardens. More planks can be added as the crack is filled higher up.

Blind exfoliation repair

For stones that are in the process of delaminating, such as face-bedded sandstone or limestone, a complex and somewhat tricky process can be used. This innovative procedure, pio-

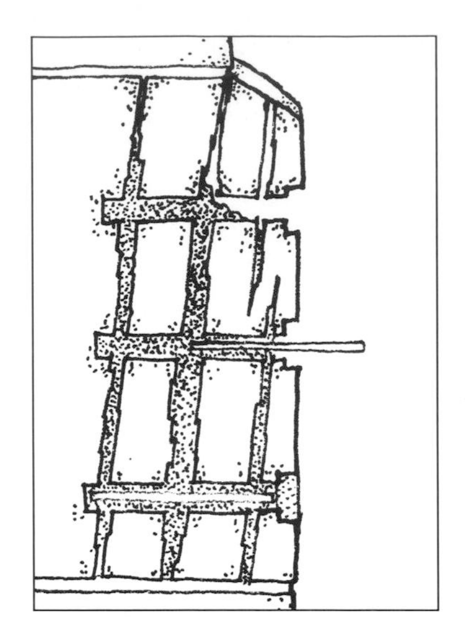

Top: Repair of delaminating stone, pioneered by the New York Landmarks Conservancy.

Above: Veneer repair with marble. The two center sections are new marble pieces. (Anne E. Grimmer)

neered by a team working for the New York Landmarks Conservancy, involves drilling staggered rows of countersunk holes through the face of the stone, injecting an adhesive grout of epoxy resin (simpler and stronger) or an acrylic-modified cement (sometimes used in larger cavities, as it is less expensive) that spreads into the gaps behind the delaminating layers, inserting stabilizing pins into the holes and finishing the surface of the holes to match the rest of the stone. A 60-milliliter, single-use syringe designed for use by veterinarians may be used.

Veneering

A new veneer of stone or a substitute material can be used to replace a deteriorated surface or to hide repair work.

If the stone must be completely removed and a like material is unavailable, the gap can be filled with mortar and a stone veneer applied to match the rest of the wall. The back of the cavity is filled with a series of thin layers of mortar (plastic repair, below) until a space exactly the size of the veneer piece is left; several days later, the sides and back of the cavity are covered with a fast-setting mortar and the veneer piece is laid in. Alternatively, the veneer piece can be mechanically attached with metal anchors without filling in behind. The veneer can be obtained by removing stones from another unobtrusive place on the building and slicing them into layers a few inches thick.

Plastic Repair or Composite Patching

Plastic repair, or composite patching, is the patching of selected areas of deteriorating stone by building up a series of coats of cementitious material to reconstruct the missing surface. The technique, used to repair problems such as delamination, exfoliation and spalling, can be quite successful if limited to small cavities or areas of missing stone (1 to 3 inches deep). If carried out by a skilled worker, plastic repair may be less obtrusive and less expensive than a replacement in natural stone. On the other hand, if poorly done, a patch not only will be visually apparent but also will shrink and separate; water will get into cracks and the whole patch can pop out, leaving the stone even more damaged.

A good patch should match the color, texture and surface

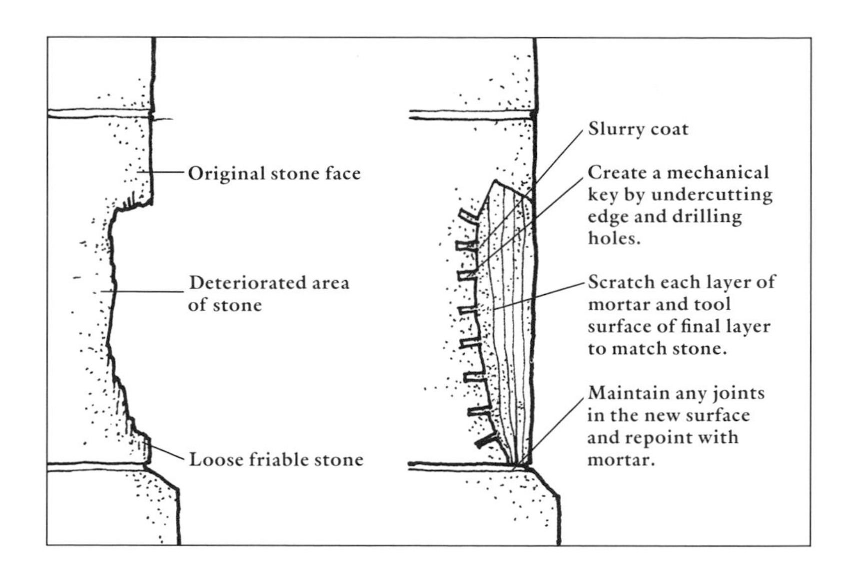

Plastic repair. A mortarlike mix is applied to the deteriorated area in layers.

treatment of the rest of the stone. It should adhere well and have physical characteristics similar to the original stone (hardness, moisture transmission, expansion and contraction) or else the existing stone will be damaged by the patch.

Plastic repair is exacting work, and only experienced craftsmen can be relied on to make a good patch. Be sure to check out previous examples of a firm's work before hiring. Even if well done, a patch will never look and weather exactly the same way as the natural stone; patches, therefore, should not be used over very large areas.

Preparation of the deteriorated masonry

The stone must be properly prepared to ensure that a good bonding surface exists between the old and new materials. The surface of the deteriorated stone is scraped and chiseled down to a sound layer; deep grooves are cut into the cleaned-out surface to provide a solid anchorage. A mechanical key is provided by undercutting the edges of the remaining stone to a slight dovetail and by drilling small holes (½-inch diameter, ½-inch depth) into the sound masonry when more than a 1-inch depth is to be filled or when work is done on a vertical surface. To make the patch fit in with the visual character of the wall, a regular shape is usually best with smooth-faced regularly cut stone and an irregular shape with rough-faced and randomly laid stone.

Each individual stone should be patched separately, otherwise a patch that crosses two or more stones will crack as

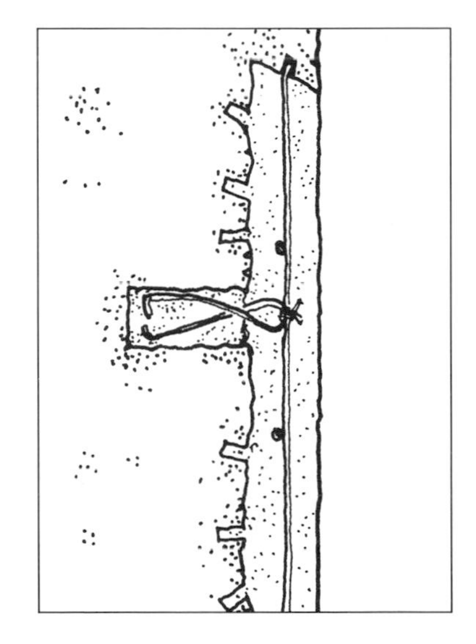

Using galvanized steel mesh with a plastic repair. The mesh is fastened to the stone with metal ties that have been imbedded in the original stone. Small holes drilled into the surface also permit better adherence of the mortar used for the patch.

stones shift. Put wood strips in adjacent mortar joints to contain the repair; the joints should be repointed afterward. A thorough washing of the area with water and bristle brushes completes the preparation.

To give better support to plastic repairs of greater depth or if the elements to be repaired have a considerable overhang (cornices, moldings), stainless-steel, copper, bronze or polyester reinforcing rods, pins, wires, screws, anchors or mesh are often used. They are sealed in place with portland cement, epoxy resin or, more rarely, lead. However, using these supplementary devices does not guarantee a successful repair.

Remodeling a molded element using a metal mesh. The mesh is bent to the original shape.

Mixes

The material used for plastic repair is similar in composition to mortar or stucco mixes: 1 part binder (lime and cement) to about 3 parts of aggregate (sand or crushed stone) made up with water. Mixes vary according to the type of masonry being repaired; they should always be a bit weaker than the masonry so that if deterioration takes place, it will be in the patch, not the stone. Plastic repairs carried out using too hard a mix may not adhere or may accelerate weathering and deterioration of the adjacent natural stone, partly because of the different rates of expansion. Epoxy should not generally be used, because it is too strong and is impermeable to water.

A typical binding agent is made up of equal parts of lime and white portland cement, with a small amount of non-reemulsifiable acrylic latex to increase cohesion and durability.

Sand and/or crushed stone is the usual aggregate. Ideally,

Test patches for a plastic repair. These are allowed to dry and harden before they are compared to the original stone. (Michael F. Lynch)

at least, the final coat should be made up with the same kind of stone as is being repaired, either crushed stone from the building itself or matching stone from a salvage or stone yard (ground to pass through a screen and washed thoroughly). If an accurate color match cannot be made with the proper choice of aggregate, small amounts of alkali-stable dry pigments can be used to make the patch blend in with the original masonry, although these may reduce the strength of the repair and the color may eventually alter.

Composite patches should also match the stone in texture and surface tooling. In comparison with natural stone, plastic repairs will look rather dull and lifeless. Natural sandstone often has mica in it, which sparkles, and a repair made with only sand will look flat; if it is not possible to get crushed sandstone for the patch, some crushed mica, glass or marble dust may be added.

Plastic repairs may, if necessary, be painted to match adjacent areas of masonry. Patches are often made to match a clean, unweathered area of the original stone (perhaps from the back of the area to be patched) so that the patch will weather to the same color; however, the color and texture of the patching material may not change over time in the same way as the natural stone. Several colors may be needed to match the natural variation between several stones or even within one stone.

The only way to get a proper color match is by preparing test samples. First, the sands, crushed stone and, if necessary, dry pigments are blended to approximate the color of the stone, noting carefully the quantity of each. Possible variations can be mixed. The dry mix gives a good indication of the final color. The mixture is completed with binder and water. After curing for 48 hours, part of each sample is treated with the proposed finishing treatment, and then samples are compared with the original stone. The process is repeated until a good match has been achieved.

Application

The mortar is laid in many thin layers, between 3/8 and 3/4 inch, to reduce shrinkage and cracking when drying. Each layer should set usually about 2 to 4 hours before the following coat is applied. The first, or slurry, coat should be the thinnest to get into the nooks and crannies of the stone. Then, while it is still moist, the first of a series of scratch coats is laid.

Top: Inappropriate composite patch. This patch does not match the color of the stone or the configuration. Ideally, each piece of stone should have been patched, then mortared. (© Richard Pieper)

Above: Oil stain on a composite patch. (© Michael Devonshire)

Restoring a deteriorated column with patching materials. (NTHP)

Each coat is wet down before the next is applied. The final finish coat should match the color, texture and surface finishing of the original stone.

To achieve the proper texture, the surface must be treated to expose the grains of stone or sand; otherwise, only the smooth surface and color of the binding material will be visible. There are several ways to do this:

- Stipple with a damp sponge or a dry trowel with a wooden float when the patch has partially cured to a leatherlike surface.
- Lightly acid-etch with acetic acid after the patch has cured at least 48 hours, ensuring that other parts of the building are well protected; great caution must be exercised near limestone and marble.
- Rub the surface of the patch after it has cured well. Surface tooling can be done either when the patch is partially cured or after it is well cured using stone tools.

Consolidation

For centuries people have sought ways to bring disintegrating (crumbling, spalling, sugaring) masonry back together and to increase masonry's strength and resistance to deterioration. The ideal agent would consolidate friable stone or brick by forming a new weather-resistant binder after it has been injected into a masonry unit. By modifying the pore structure, which would reduce the absorption of water and other harmful substances, and by increasing tensile strength, it would lessen damage caused by salt crystallization. It would have good penetration beyond the depth of subflorescence (generally between ½ to 1 inch) to the undamaged core, would create no harmful byproducts, would allow the masonry to continue to breathe and would not change its color. Thus far, this miracle formula has not been discovered.

Materials

A variety of substances have been used with limited success. Limewater, a clear saturated solution of lime in water, was commonly applied to limestone, particularly in Great Britain, as a kind of natural consolidant; oils, waxes and resins were also used. Generally, these substances did not penetrate into the material and behaved as coatings, with all their related problems (page 138).

In the past 20 years, experiments with synthetic consolidants such as silanes, acrylics and epoxies have resulted in treatments that penetrate deep to coat pores within the stone; they appear to avoid many of the problems of surface treatments but they are expensive and irreversible.

Use

Although some of the recent experiments using synthetic consolidants, carried out mainly in Europe, look promising in terms of short-term effects, several issues must be resolved before consolidation can be recommended for general use. The application of these products is not reversible; if something goes wrong, there is no way to remove them. Also, the life of a treatment is not known, nor is it clear whether it would be possible to reapply another treatment or how many treatments could be applied before pores are totally blocked. Therefore, consolidation must still be considered an experimental technique, too risky for general use and also precluded by its high cost.

Consolidant applied to sugaring marble. The consolidated marble is at the bottom. (Christina Henry)

For now, the only situation that might warrant consolidation would be stone with important detail that is deteriorating so rapidly the material would be lost in the next few years. Even for this rare exception, consolidants should never be used without comprehensive laboratory testing and field evaluation and the advice of an experienced, independent stone conservator knowledgeable about these experimental techniques.

Now that these systems are gradually being introduced into the North American market and with scientific research continuing, the level of expertise on consolidants should be greatly improved. It may well be possible in the near future to select a good consolidant appropriate for a given stone suffering from a particular problem. Therefore, for stone that is weak and friable, but not on the verge of total failure or of damaging the building, it may be wise simply to wait until these consolidation techniques are perfected, rather than cover the masonry or use some of the other techniques described in this chapter.

Brick and Stone Replacement

If a brick or stone is so badly damaged that it cannot be repaired, it must be replaced. While the following focuses primarily on stone, some of these techniques apply to brick as well. For a more detailed discussion on replacing brick and repointing brick and stone, see Repointing Brick and Stone.

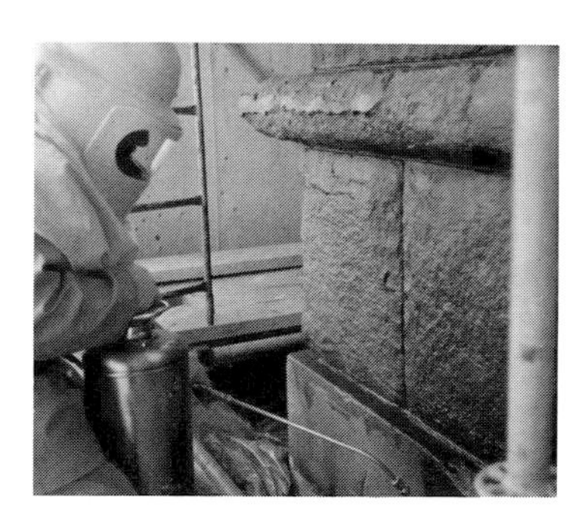

Consolidation of sandstone. The structure (top, Royal Bank of Canada) is sheathed in plastic to protect passersby and maintain warmth as work progresses. Hazardous chemicals make protective clothing mandatory (center and above, ProSoCo, Inc.).

Replacement Material

It is always best to replace a brick or stone with an exact ıatch of the original in terms of composition, color, texture, rength, finishing and porosity. Even with stones or bricks ithin a wall that may not be seen, a match is preferable at ast in terms of size, strength and moisture resistance.

The best match for a stone is most likely one from the me quarry. A conservation architect or stone distributor ıould be able to identify a local stone and determine where it was quarried. If the original quarry is no longer in operation, the equivalent may be obtained from another quarry. Sometimes matching stones may be salvaged from a demolition yard. Failing that, an existing stone from the building can be sliced and used as surface veneer to replace several missing stones (page 148).

Ensure that the new stone is of good quality with no clay veins or other obvious weaknesses. In the case of sedimentary rocks, be sure that the block is cut so that the bedding planes will be in the proper direction.

The exterior face of a replacement stone should be finished with the same surface treatment as the original, and all ornamental details should be reproduced.

Artificial stone may be used to replace stone, matching the surface finish to a certain extent in the casting process. Pigments may also be used to match the original stone; however, artificial stones tend to age differently from the surrounding masonry. These concrete units are often reinforced with steel, as would be the case with modern precast panels, particularly if they play a structural role as in lintels. The reinforcing bars should be made of stainless steel or at least be well set in from the surface (1 inch); otherwise, they are prone to water penetration, which would lead to rust and cause the concrete to crack or the surface to spall.

Because brick is a manufactured product, it may be difficult in today's market to find a brick replacement that matches the original's dimensions, color, finish and shape. In some cases, it may be necessary to order custom-made brick from a manufacturer or craftsman who specializes in reproducing historic brick.

Old bricks may be found in a demolition yard or site or bought from a dealer who specializes in recycled building materials. Cracked or damaged bricks should be avoided, and lower-quality, more porous common or fill bricks should never be used as exterior facing. Because it is often difficult to identify a face brick by sight, testing may be required. Before using

Replacing a stone.

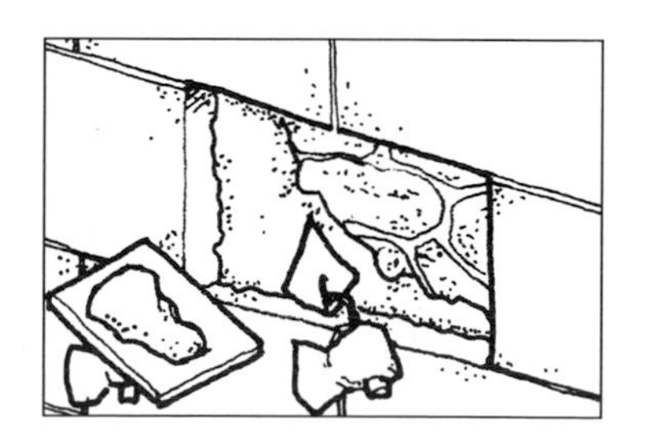

second-hand bricks, all of the old mortar must be removed, not an easy task as the brick is easily damaged, particularly if the mortar to be removed is cement.

Even with a relatively good match, the area of replaced brick and stone will probably look slightly different from the rest of the wall, at least until the replacement units have weathered. To prevent the wall from looking too mottled, it is best to make the area as simple a geometric shape as possible.

Installation

If only a few isolated stones or bricks must be replaced, each should be removed without touching the surrounding masonry. When the deteriorated stone or brick is chiselled out, care should be taken to avoid damaging the edges of the adjoining elements, particularly when removing the old mortar.

The replacement stone or brick can then be slid into the space left by the old. The fit of the new element should first be tested by fitting it into its space without mortar. A stone is placed on wedges of wet wood (or if it is a heavy block, of noncorrosive metal); the wedges are positioned so that they will be covered with at least 1½ inches of mortar when pointing is completed. Any necessary corrections can be made at this stage.

The four sides of the space and the back face, if it is a solid masonry wall, are covered with sufficient mortar so that there will be no air spaces left when the replacement element is pushed into place. The new stone is lined up and properly set up by tapping it with a wood or rubber mallet; if it enters too far, it must be removed and more mortar must be added in the space. With a large stone it may not be possible to rely on mortar to keep the new stone in place, and it may be necessary to fit it in place with metal anchors. A brick is laid on a bed of mortar that matches the depth of the surrounding horizontal joints.

After all the new elements have been installed and adjusted, the joints should be repointed to match the rest of the wall. The repaired area should then be scrubbed clean with a brush and water to remove all excess mortar. Finally, in some instances, it may be desirable to weather the surface of the stone artificially.

Preserving Original Materials

In all rehabilitation work, preserving the building's original materials is the most important goal. Here are the measures to consider with the advice of an expert. Beginning with the simplest and least intrusive repair, each item increases in complexity.

1. Secure loose pieces of stone before they separate; reattach separated pieces whenever possible.

2. Leave the material as is if there is only slight damage.

3. In thick stone with no decorative detailing, cut back delaminating layers to sound stone and retool the surface to match the existing wall.

4. Install a matching piece of natural or artificial stone, called a dutchman, if part of the surface is missing (page 147).

5. Fill in a deteriorated section of between 1 and 4 inches deep with a composite patch (page 148).

6. Remove delaminating stones that have little significant surface texture or detailing and reuse them, perhaps by reversing and replacing them on the surface. Some retooling may be necessary.

7. Totally replace an area (veneering, page 148) or a whole stone (page 154) only when the entire surface is missing to a depth of more than 4 inches.

All the methods described here depend on a high level of skill. Only specially trained and experienced craftsmen should be employed, they should be closely supervised, and the process should be well tested on a sample area on an unobtrusive part of the building.

Honing a deteriorated stone surface, work to be done only by experts. (Pietra Dura, Inc.)

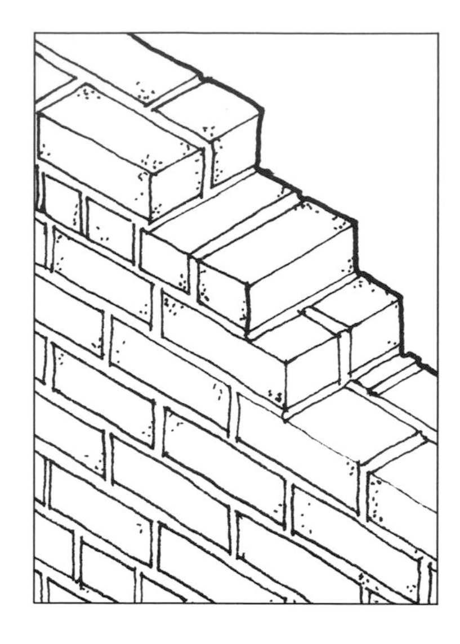

Top: Solid brick wall. The wall is tied together with a row of headers every sixth or seventh course.

Above: Solid stone wall. The wall comprises two facings of stone separated by an infill of broken stones. Wood members integrated with the inside wall provide an interior nailing surface.

Building Parts

The first section of this chapter dealt with the deterioration and repair of the masonry material itself. This section deals with the problems and repair of assemblies of bricks and stones in various parts of a building.

Walls

Until the second half of the 19th century, most brick and stone walls were solid and load bearing, supporting the framing of the floors and the roof. The thickness of a solid wall was a function of the loads it had to carry, its height and its construction material. Early stone walls were made of fieldstone. Later, exterior and interior walls were faced with cut stone, with the space between filled with rubble (small stones or quarry debris and mortar). Through-stones or headers (large stones that extend the full thickness of the wall or penetrate deeply into it) tied the wall together and prevented it from buckling. In the mid-19th century the use of brick infill behind the cut stone exterior facings became common. Solid brick walls of houses and small buildings were usually two or three bricks or wythes thick; the walls of larger buildings could be thicker, as much as four wythes or 16 inches thick. Many different bonding patterns were developed to knit the wall into a single structural entity with header bricks placed to span from one wythe to the next one.

In the late 19th century the solid wall gave way to the cavity wall, with two or more wythes separated by a continuous air space. The use of masonry cavity-wall construction had been common in Europe, particularly in England, and had even been used by the early Romans. The inner wythe is usually load bearing; the outer one, the facing, might support itself but rarely supports other parts of the building. In the 20th century the structural support for the building is often provided by steel or concrete columns and neither the inner nor the outer wall is load bearing.

Originally, both wythes were built of stone or brick. Later, the interior load-bearing wall was made of other materials such as wood or concrete block. The masonry veneer was anchored to the inner wall with masonry headers or metal ties, first simply with long nails driven partially through the masonry into the wood, then with specially manufactured metal ties. Metal ties should be corrosion resistant and sufficiently flexible to absorb the movements of the structure. With brick facing, they were usually placed every sixth course, or about 32 inches apart.

A cavity wall usually can be distinguished from a solid bearing wall by the

- Absence of headers. (Sometimes, however, for decorative purposes, half bricks were used to give the illusion of headers.)
- Thickness
- Age of the building. (Cavity walls became widespread in the 20th century.)
- Presence of a crack on the inside wall paralleling one on the outside. (This is an indication that it is a solid wall.)
- Presence of weep holes. (These are vertical unfilled joints at the top and bottom of the wall and over the lintels, left open to allow moisture to escape from the cavity.)

The space between the two walls reduces energy loss through the wall by limiting thermal conductivity. It also prevents water and vapor from penetrating inside the building by providing a space within the wall from which moisture can be vented; thus, interior finishes run less risk of damage. To be effective, the cavity must be continuous, properly vented and at least 1 inch wide.

Condensed water that drips to the bottom of the air cavity should be prevented from penetrating into the foundations or openings through the use of metal or vinyl flashing, installed at the bottom of the wall or above lintels. The flashing should extend under the masonry facing and overhang to form a drip edge. The wood framing should be protected from moisture with building paper.

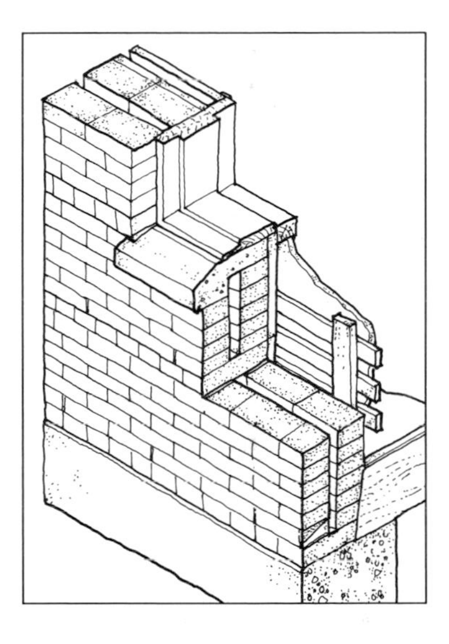

Cavity wall. Two layers of masonry, usually brick, are separated by an air space. The inner wall is generally load bearing; the outer wall acts as a facing and often includes weep holes for ventilation. The two walls are bonded together with headers or metal ties.

Insulation

High energy costs along with government programs of the past decade have spurred a trend to upgrade the insulation of buildings. It is very difficult, however, to add insulation to an existing brick or stone wall without completely destroying or covering either the interior or exterior surface. The economic benefits may be marginal, especially if other insulation and weatherstripping have already been installed. Also, the long-term effects of high levels of insulation in old buildings are not yet known; of particular concern is the effect of sealing an old building to prevent the movement of air, often resulting in excessive condensation.

If the interior finishes of a structure must be replaced, the best way to insulate is to add insulation on the inside face of the wall. Spraying insulation into the air space of a cavity wall is especially problematic. It is hard to reach all parts of the space, and filling the void defeats the purpose behind the cavity wall — a ventilated gap between the inner and outer wall that promotes the natural evaporation of moisture.

Whenever insulation is added to a wall, an air or vapor barrier should be added on the warm side to prevent moisture from condensing within the insulation.

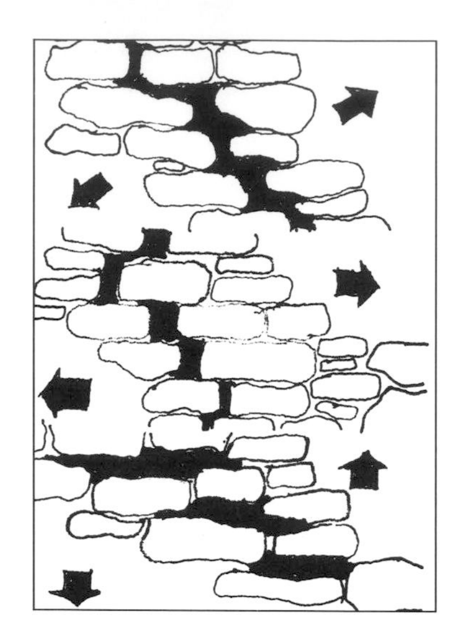

Cracks in a wall. Examining the shape and direction of cracks will help determine their origin.

Cracks in walls

Once their source has been identified and rectified, small cracks can be filled without disturbing the adjacent stone or brick. First, all of the crumbled or deteriorated masonry on

Repairing a stucco wall. (Dinu Bumbaru)

either side of the crack must be removed, and the crack itself should be scrubbed clean with a hard masonry brush.

The crack then is sealed with a lime and cement grout that matches the color and texture of the wall. To make the repair more discrete, the color of the mortar can be adjusted by adding stone dust and pigments and the mortar can be textured with various tools (trowel, brush, finishing tool).

Grout, a watery mortar, is generally sufficient for a crack up to 1 inch wide. With wider cracks, small holes are sometimes drilled into the sound stone to increase the bonding area. Reinforcing mesh or clamps can be fixed to the masonry to support the grout; the mesh should be set back at least ½ to ¾ inch from the finished face of the wall so that it will be well covered with mortar. Using a small, tightly fitting formwork attached to the surface of the crack, the grout can be injected into the crack and held in place until it sets.

The corners of a building or the place where an addition meets the original building are particularly susceptible to cracking because the two walls were built and may behave differently. If rebuilding is necessary, a specialist may recommend installing an expansion joint, a vertical gap filled with a flexible material that allows movement to take place without cracking adjacent masonry walls.

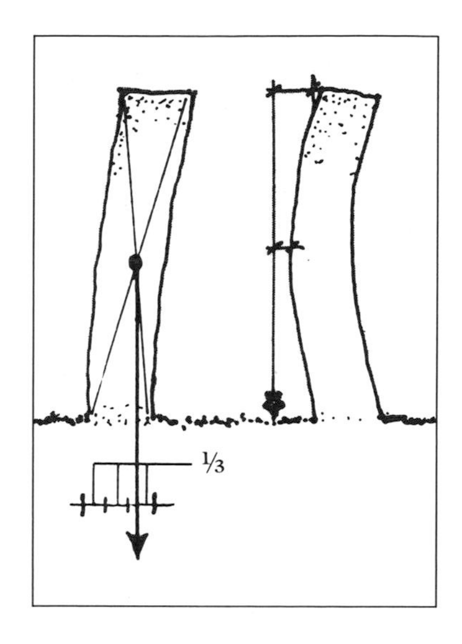

Wall deformations. If the wall is tilting (left) and the center of gravity is outside the middle third of the wall, the structure may collapse. The extent of bulges (right) can be measured with a plumb line. The problem may stem from settlement or poor masonry bonding within the wall.

Bulges and tilting

Deformation in a wall may be the result of major structural problems with the building such as settlement of the soil, inadequate foundations or the deterioration of the wood structure in a cavity wall. Or the structure may be stable, but the outer facing is separating from the rest of the wall because it is not well attached. This condition results from insufficient or poor distribution of headers, the bricks and stones that tie the wythes together, or from the deterioration of metal ties in a cavity wall exacerbated by the expansion of water freezing in the wall.

A commonly used repair technique for walls that are bulging or tilting is to secure the masonry with steel tie rods. These are solidly fixed to the floor beams or the inner wall of a cavity wall, piercing the masonry wall. Attached to S hooks or metal plates such as stars on the exterior, the rods hold the masonry in place. This solution, however, may lead to other problems. Tie rods create holes in the wall through which

S hook, used with a tie rod to anchor floor beams to the exterior wall and to minimize wall bulges and tilting.

Wall reconstruction. This wall was beyond repair and required replacement. Wooden shoring can temporarily keep a damaged wall in place until it can be repaired. (Dinu Bumbaru)

water may seep. In cold climates, they become "cold bridges" — short cuts for the cold to reach the more humid interior. Moisture condenses on the metal rods, penetrating nearby wood, which can then rot. Also, new metal plates are not particularly pleasing aesthetically, although traditional decorative cast-iron plates sometimes are part of the character of commercial and industrial buildings.

Repairing or rebuilding a solid wall

A solid, load-bearing wall is relatively difficult to repair. When a large part of a solid wall is defective, it is often better to take it apart and rebuild it. Temporary support of the structure may be necessary to accomplish the job.

Often, the rubble within a stone wall has so deteriorated — sometimes as a result of moisture washing away the mortar used with the rubble — that it has lost all cohesion and the facing stone is detaching. In this case, it may be possible to inject grout, which will reconsolidate the wall in place in a process called pressure grouting.

If the inner part of the wall is sound and the problem is only in the facing, it may be possible to repair only the latter. The inner wall can be given added support by anchoring it to new studs attached to the inside face of the wall. The studs can also be used to install insulation. Again, temporary shoring may be required.

Most likely, however, a deteriorated wall with rubble fill will need rebuilding. An accurate survey of the masonry and the placement of the decorative elements should be done before dismantling. Each reusable cut stone should be numbered with chalk on a hidden face so that it can be reinstalled in its original location during reconstruction. The stones must be handled and stored carefully and should not be stacked on top of one another. When the outer face has been removed, the exposed interior wall should be protected from rain with a polyethylene sheet.

With brick it is more difficult to reuse the original material. Cleaning off the old mortar is painstaking and costly for an entire wall. Nevertheless, the brick pattern should be carefully noted so that the new facing follows the original brick motifs — often the only decorative detailing on a building. Bricks with special finishes (for example, glazed) or shapes (curved, molded) should be cleaned and reinstalled.

When the facing stone or brick is relaid, additional

strength between the facing stones and interior wall can be provided with stainless-steel straps or wires. With brick, galvanized metal ties are secured into the fill brick (a better practice than merely inserting them into hollowed-out mortar joints) and then embedded at regular intervals in the joints of the new facing.

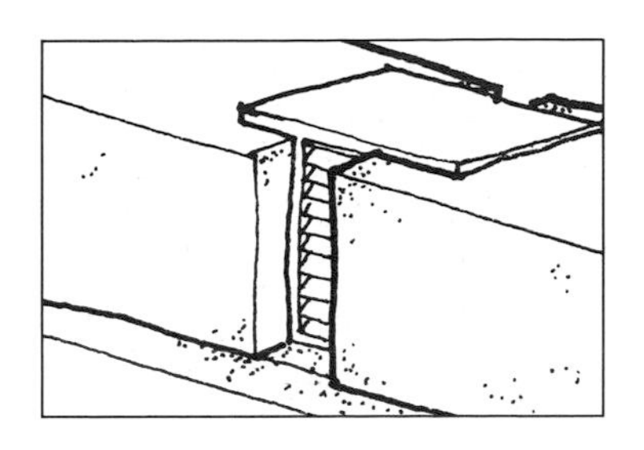

Prefabricated plastic weep hole. This device prevents weep holes from being filled with mortar as a wall is reconstructed.

Repairing a cavity wall

Because the outer face of a cavity wall generally is not load bearing, it is usually easier to repair than a solid wall. Although dismantling the facing of a cavity wall normally does not call for shoring, repairs on the lower part of the wall require support of the upper part with wood members or steel jacks. Alternatively, the masonry can be dismantled in the shape of a triangle (such as above a window or door opening) in such a way that the remaining part of the facing continues to support the upper portion. After dismantling the brick or stone, any rotted wood — generally at the bottom of the wall where moisture accumulates — should be replaced. If the inner structure is wood, it should be covered with building paper, tar-impregnated paper that limits moisture and air movement. Install galvanized metal or vinyl flashing at the base of the wall and above the openings. Galvanized ties are nailed at regular intervals, generally at every 16 inches vertically and every 24 inches horizontally.

The stones or bricks are laid in a mortar bed in which the metal ties fixed to the inner wall are embedded. The joints are then packed and repointed. Care must be taken not to drop any mortar in the air space, as it may block weep holes or accumulate at the bottom of the cavity, where it may retain moisture and lead to rot.

Entranceway. Heavy traffic use creates special and sometimes severe problems for entrances. (Dinu Bumbaru)

Window and Door Openings

Openings in a wall such as doors and windows are composed of a sill at the bottom, a jamb on each side and an arch or lintel at the top.

The design of doors and windows generally plays an important part in the architectural treatment of the building and has changed with different periods and styles. Often better materials were used around openings to accent the design. For example, in rough stone or brick walls cut stone was some-

Brick segmental and flat arches. An arch's rise influences solidity. There should be a rise of at least 1 inch for each 12 inches across.

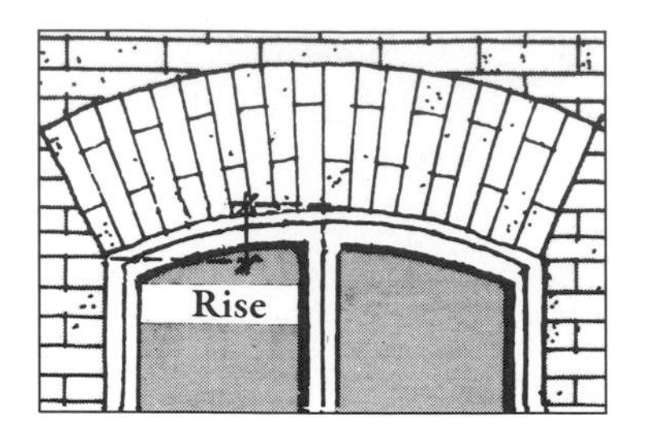

times used to surround openings, as sills or lintels, or for the keystone or springing element of an arch.

With a bearing wall, the arch or lintel transfers the weight of the masonry above the opening to the abutments and to a certain extent to the wall between the openings.

Arches

An arch is the only means of spanning a large opening with stone or brick, which are weak in tension but strong in compression. Each stone or brick in an arch is called a voussoir. Arches are built on site, on wood or metal formwork, which is removed once the final voussoir, the keystone, is in place.

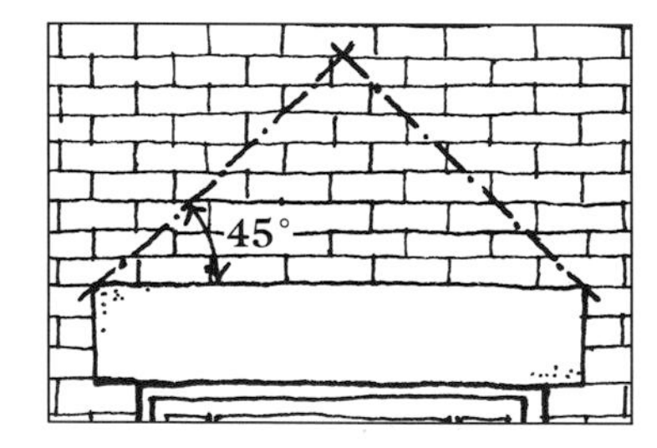

Lintels. Like arches, lintels carry the weight of the masonry above. Cracks mean either weak lintel material or a poorly designed arch.

Lintels

A lintel is essentially a small beam over an opening. Made from a single piece of stone, wood or, more recently, steel or concrete, it supports the weight of the wall above it. A stone lintel is subject to tension stresses and, therefore, risks cracking if the span is too wide for the depth of the lintel.

In cavity walls, the air space is usually interrupted by the lintel or the steel angle that supports it. If this is the case, flashing should be inserted above the lintel and weep holes should be provided to drain water that may collect.

Sills

A sill covers the bottom of the window or door opening and prevents water from penetrating into the walls. It is generally made of a single piece of stone, sometimes wood or, recently, concrete.

A sill must have a slope of at least 10 degrees to drain, and it should also have a "drip," a small groove found on the underside edge of the sill. This groove prevents water from running back under the sill and onto the wall, where it can lead to deterioration of the mortar, the masonry and, eventually, the wall of the building itself. For the same reason, it is also important for the sill to project sufficiently from the wall.

Problems and repairs

Structural problems with buildings often manifest themselves at window and door openings first and can result from inherent weaknesses in the construction of the opening or from more general structural problems. Inherent defects may include

Filled-in arch, an inappropriate repair.

- Stone lintels that are too small or not well supported. (The triangular section of masonry above the lintel can give way.)
- Arches or flat arches with insufficient rise. (Dislocation can result in opening of the joints and movement or collapse of the arch.)
- Abutments and lateral walls too weak to support the horizontal loads transferred from the arches. (This can lead to cracking in the walls.)

To remedy such problems, the defective element must be reinforced or taken down and rebuilt properly. Before dismantling, the wall over the opening must be shored up by inserting a steel angle in the joint immediately above the lintel or arch.

Cracked lintels can be reglued or replaced. If the lintel is too narrow or shallow, it can be reinforced with steel angles laid on the abutments or attached to the structural framing of the building. In some cases, the situation may have stabilized, and the lintel or sill could be left as it is provided that a sealant is set in the crack to prevent water penetration, but weep holes must be provided for ventilation or the moisture will be trapped. Arches may need only to be rebuilt or may require reinforcement with curved metal angles resting securely on the abutments or the structure of the building. If the sill has no drip, one can be cut using a radial saw. Although a skilled workman might be able to do this in place, it is generally necessary to remove and reinstall the sill.

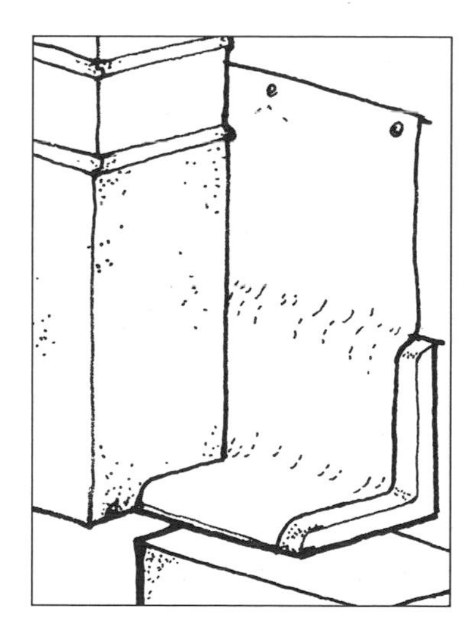

Steel angle reinforcing a weak lintel. Flashing has been installed above the opening.

An insufficient slope or the presence of cracks in the sill are also causes of water infiltration. The sill must then be repaired and reinstalled or replaced. When doing this, flashing can be placed under the sill to reduce the chance of water infiltration. Flashing can also be added when replacing a lintel in a cavity wall.

Parged parapets with copings. (© Michael Devonshire)

Parapet showing possible paths of water penetration. (Forrest Wilson)

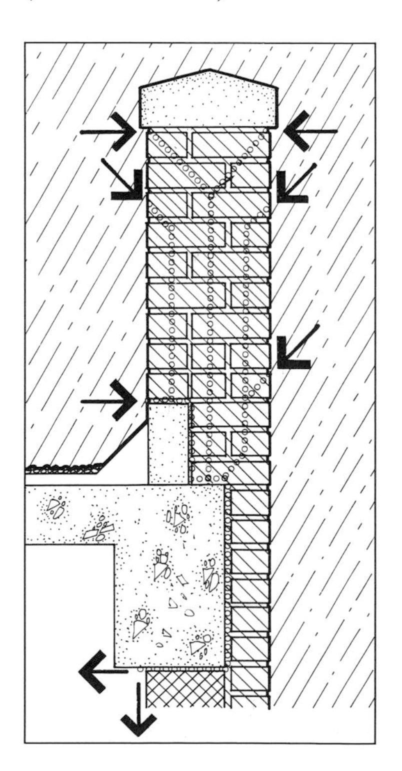

Parapets and Cornices

The parapet, the top part of a wall that extends above the roof, is often the weakest point in a building's protection system. The junction between the roof and wall is located in a highly exposed but rarely inspected location. Water often accumulates nearby, both the front and back faces are exposed to the elements, and there is no heat in the wall to help dry out moisture.

To protect the top of a wall against the infiltration of water, the parapet is capped with stone, terra cotta or concrete slabs or metal copings, which are gently sloped to drain. Copings should overhang the wall to keep water away and should have a drip.

Buildings with flat roofs, which generally have cavity walls, usually have a metal coping. With low parapets this coping interlocks with the roof's counter flashing and is nailed to a wood strip incorporated into the facing masonry. Even if the coping is well sloped toward the roof to drain off water, rust holes are always a possibility, so the roof membrane should be extended under the coping.

Traditionally, copings extend at least a few inches out from the wall; this is quite effective in reducing weathering on the top few feet of the wall. In new construction or renovation, there is a tendency to include a lip on the coping that extends out by a fraction of an inch; a coping of at least 4 inches should be provided.

Cornices are projecting horizontal elements that crown or subdivide facades or decorate the lintels over openings. In solid walls they are made from header stones, well anchored into the masonry wall. For cavity walls or large overhangs, they are anchored to the structure with metal ties.

Because a cornice projects from the facade, it must be designed to shed water. The top face should have a slope of at least 10 degrees; current construction and rehabilitation practices call for the tops of cornices to be covered with metal flashing. The flashing should be similar in color to the material it is covering and should cover the whole top of the cornice. To protect the joint between the cornice and the masonry facing, bring the flashing up to the facing and interlock it with the parapet coping or insert it into a raked joint in the wall.

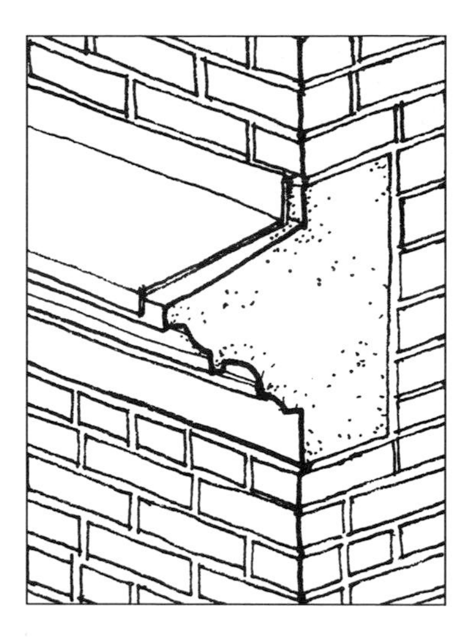

Cornice flashing imbedded in masonry. Lead-covered copper flashing can be used.

Applying new stucco to a cornice at Decatur House (1819), Washington, D.C. Two layers of stucco were placed directly on the brick without a wire lath, as had been done historically. (David Baxter)

Work completed, after a coat of paint. (David Baxter)

Gable and chimney of the Glessner House (1887), Chicago. (HABS)

Chimneys

The chimney, which has its own foundation to carry its concentrated weight, includes the fireplace, flue and stack that extends above the roof. Until the recent use of metal, chimneys had to be made of masonry, even in wood buildings. In traditional houses, they were usually located at the center of the building or on a gable wall and were often built within the end wall. In row housing, they were typically built within the party wall.

Today's building codes usually contain detailed requirements to reduce fire hazards. These may affect how a chimney is repaired or rebuilt, including the type of noncombustible materials used, the size of the flue and the hearth, and the height of the stack above the roof ridge. Fireplace dimensions vary, depending on the fuel — gas, coal or wood. A fireplace designed for gas is probably too small for burning wood.

The back plate and the hearth of the fireplace must be fireproof; in old buildings, they were made of stone, firebrick or tile, often surrounded with a small relieving arch to simplify replacement of damaged bricks.

The flue links the fireplace to the stack and includes a

Bracing for a chimney. Bracing is tied to the roof with a plate inserted under the roofing material.

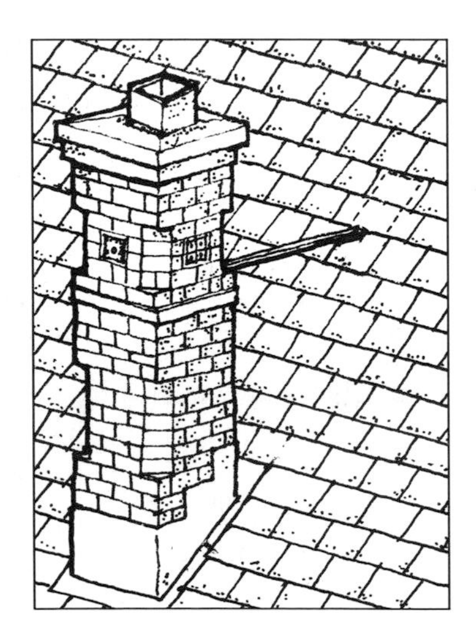

smoke chamber that prevents downdrafts from filling the room with smoke. The flue was made from rough stone or brick, sometimes laid with clay mortar. The inside was generally lined with terra cotta or refractory brick or, more recently, cement or metal. A flue that was not well lined or that has a deteriorated lining may cause a fire. The flue also may be blocked or damaged by combustion residue or condensation resulting from certain types of heating material such as gas or softwoods.

Chimney flues should be swept regularly to prevent the accumulation of soot and creosote, which may eventually lead to a fire. A severely damaged chimney may be dismantled and rebuilt — an expensive process requiring specialized workers. Stainless-steel or new terra cotta liners may be installed to replace the flue.

Because stacks are exposed to the weather and rarely have caps, water infiltrates and leads to faster deterioration than in the rest of the chimney. Today, caps made of bevelled mortar or a stone slab are installed to project at least 2 inches beyond the face of the stack. A drip sheds the water away from the stack.

The stack may deform with time, often as a result of chemical reactions among the mortar, smoke, air pollution and moisture. If the stack is leaning, it may be necessary to have it dismantled and rebuilt, preferably in the original design. The chimney stack can be braced with a metal rod attached to the roof; the rod is connected to a plate inserted under the roofing in a such way that water infiltration is avoided.

Stone chimney cap. The cap could also be made with mortar or brick, although stone is the preferred choice.

Chimney reconstruction at Oatlands. The chimney dismantled (left) and rebuilt. (David Baxter)

Rubble foundation.

Brick foundation.

Reinforced-concrete foundation with footing.

FOUNDATIONS

A foundation is the base of a wall. It serves to spread the weight of the building on the ground, keep out ground moisture, resist the pressure of the earth pushing inward toward a basement and, in cold climates, anchor the building to a stable part of the ground below the depth where it heaves because of frost.

The design of foundations is determined by the weight of the building it supports, the type of soil it rests on and the climate, which also determines the depth at which the ground is free of frost.

The footing provides a seat for the foundation wall; distributing the loads to the ground over a larger surface, it helps prevent settlement. In northern climates, the footing lies below the frost line. A foundation that is too shallow may heave with frost, causing cracks or settling in the building.

Evolution of foundations

With early construction in North America, buildings such as log cabins and simple wooden houses often sat on simple foundations of a few layers of fieldstones, laid without mortar and perhaps resting on a row of larger stones.

While this type continued to be used in outbuildings, foundations of main buildings gradually changed to thick walls of roughly squared rubble, assembled with a lime or, later, a cement mortar. These walls were often so thick that they did not need footings, or they sat on broad stone footings.

In the 19th century cut stone was sometimes used for the part of the foundation above the ground, and brick foundations were also common.

Poured foundations appeared at the end of the 19th century. Planks were laid along the sides of a trench, stopping short of the bottom so that the cement would spread to fill up the bottom of the trench — to create a spread footing. Cement was poured into the formwork along with stones and small rubble as aggregate.

By the turn of the 20th century, reinforced concrete became the most common material for foundations, although in some regions the use of concrete block is common for small buildings. A reinforced-concrete foundation has the advantage of acting as a monolithic whole; it thus resists moisture penetration better than foundations made of small units with many joints.

Problems and repairs

One of the main problems related to foundations is moisture penetration. This leads not only to an unpleasantly damp basement, but also to serious deterioration of the structure. Various techniques of controlling moisture (drainage, parging, waterproof barriers) are discussed beginning on page 185.

Insulating. If excavation is required to install a drain or waterproofing outside the foundation wall, it may be a good

Repairing a foundation wall at Chesterwood (1898), Stockbridge, Mass. Heaving from frost and moisture led to deterioration in a foundation wall at the Chesterwood Studio (above, Chesterwood Museum Archives). To reconstruct and strengthen the foundation, a concrete wall was installed and faced with the original foundation stone. Formwork (left, Restoration Workshop) was constructed for concrete.

Poured concrete in place, hardening. (Restoration Workshop)

Resetting the Chesterwood Studio's stone foundation wall. (Restoration Workshop)

Repositioning the head-stone. (Restoration Workshop)

Stones in place. The only job remaining is to fill in the area adjoining the wall with earth and drainage material. (Restoration Workshop)

opportunity to add insulation on the exterior. It should be installed only below grade and stop short of the footings. By insulating the exterior, the foundation wall is kept warm and is not exposed to extreme fluctuations in temperature and frost action. The insulation is protected with parging, reinforced with a galvanized steel mesh.

Underpinning. Underpinning is the strengthening or rebuilding of foundation walls and is carried out to repair or replace damaged sections, pour new footings under the existing wall or repair cracks in the wall. It may also be done to extend the wall downward below the frost line or to create or enlarge a basement. Underpinning should be done only under the direction of an engineer who can accurately calculate the structural design.

Shoring supporting floor joists during underpinning work. The shoring is made up of a steel jack resting on a footing of wood members. The jack supports a wood beam, which carries the load of several joists.

Before underpinning, the wall must be temporarily relieved of its load. Joists can be supported on temporary beams, which in turn are held up with steel jacks or wood posts resting on temporary footings, such as 4-by-4–inch pieces of wood stacked on well-packed ground. Another technique involves the use of steel beams, which go through the wall and rest on steel columns or jacks supported on temporary footings.

Once the shoring is in place, the foundation can be repaired or removed. The shoring is kept in place until the work is completed and the mortar or concrete has set.

A technique for carrying out major foundation repairs or replacement without shoring is to work in alternate sections (usually about 3 feet) so that the remaining sections continue to support the structure above.

In weak or sandy soil, an experimental new technique being developed to deal with settlement problems involves boring a large number of holes through the wall and deep into the ground, ideally down to bedrock. These holes are injected with concrete to form a web of additional support.

Stairs

Exterior stone stairs are often found on public and commercial buildings, large houses and row houses. Sometimes only a single block of stone is used as the first step to keep wood or metal stairs off the ground, thus reducing the risk of rot or rust.

Stabilizing a double stairway at Drayton Hall (1742), Charleston, S.C. Posing a hazard to visitors, the riverfront steps (top right) of this historic property were dismantled (top left) and reconstructed (above right) using as much of the original stone (above) as possible. Original stone pieces were numbered for accurate repositioning. Replacement stone was carved from an Indiana limestone that matched the original English Portland limestone. (Drayton Hall)

A stair should be made of a very hard, fine-grained stone to withstand the effects of water and excessive wear. The bedding planes should be vertical and perpendicular to the front edge so that the edges of the masonry are exposed to wear.

Each stair tread is simply a long slab of stone that rests on the stair below it and supports the one above. Sometimes, the treads rest on a rubble infill or, more rarely, on a half-arch that is supported on the building. A tread may have a straight or bullnose (rounded) nosing on its front edge and should slope slightly forward for drainage.

The deterioration of steps is due largely to their heavy use, constant exposure to the weather, inadequate support and, as independent structures, shifting with respect to the building. The stone can be attacked by de-icing salt and can also be damaged by overzealous chopping of ice.

Repairing damaged stair stones consists of variations of the mechanical repair and composite patching techniques described on pages 145–52. Alternatively, the broken stone can be replaced by laying a new stone on a mortar bed. Simply adding a new cement coating on old steps is not a durable solution, because water inevitably infiltrates between the two materials and causes the cement to become detached.

Badly damaged stairways must be taken apart and rebuilt.

Construction of exterior paving. The path is cambered to permit proper drainage. Pavers are laid on a 1-inch bed of sand or fine gravel laid in turn on a 6-inch bed of fine crushed stone.

Pavers

Paving stones and bricks are generally laid on a layer of tamped sand or packed gravel; in warmer climates, they may also be laid on a mortar bed with mortar joints, although these have a tendency to crack.

The stones used outside are generally granite, hard limestone, sandstone and slate. Because slate tends to split under frost action, it should be avoided in northern climates. The thickness of paving stones varies between 2½ and 4 inches. Stones may be used in large slabs or small cobbles.

For brick pavers, only very hard, long-fired paving bricks should be used, especially in locations of extreme weather and heavy traffic; common bricks or even face bricks should be avoided.

A wide range of cement paving bricks (unit pavers) is now available, including many traditional shapes.

Damaged exterior paving is generally caused by soil movement or poor bonding. The paving should be removed and the base under the paving examined. If necessary, the bed should be dug out and a new bed built up, with, for example, 6 inches of crushed stone topped with 1 inch of well-tamped sand. The paved area should slope slightly for drainage.

Brick pavers. (Jack E. Boucher, HABS)

Finding and Treating Moisture Problems

Moisture is probably the greatest source of damage to old buildings and is perhaps the most difficult to identify and correct. Excessive moisture attributable to rain, ground water and condensation can inflict damage ranging from dampened wallpaper and plaster to severe deterioration of structural components such as the rotting of wood and the rusting of metal. Moisture can jeopardize brick and stone walls and threaten building stability.

Many moisture problems can be diagnosed and treated by experienced architects and other building professionals applying the diagnostic and treatment techniques discussed here and explained more fully in *Moisture Problems in Historic Masonry Walls: Diagnosis and Treatment* by Baird M. Smith, AIA, on which much of this chapter is based (page 197). But even highly trained experts may well be unable to come up with a successful treatment for some situations.

Possible solutions for moisture problems include reducing the presence of moisture with better drainage, damp-proof courses (physical and other barriers that prevent moisture from rising up a wall) and, on rare occasions, surface treatments. These should be followed, if necessary, with treatments to remove salt encrustations on or within the masonry, which build up as a result of the moisture problem.

Evidence of moisture problems in stains on brick and deteriorated flashing and gutters. (Baird M. Smith)

Opposite: Frank Lloyd Wright's Fallingwater (1936), Mill Run, Pa., a concrete and stone International Style landmark subject to moisture problems. (Western Pennsylvania Conservancy)

Sources of Moisture

Moisture is always present to some degree in building materials. The point at which moisture can be seen or felt or when it can damage a building varies from one material to the next.

Moisture must generally be introduced into the material from a source such as rain, ground water or condensation.

Building design defects and poor maintenance will exacerbate problems; very often deterioration or poor workmanship in building elements such as flashing, roofs, windows and foundations, rather than the masonry walls themselves, causes the problems.

Rain

Light rains followed by periods of sun normally do not cause direct moisture problems for masonry walls. Rain and moisture penetrate into brick and stone to varying extents, then evaporate during a period of dryness. On the other hand, heavy or wind-driven rains force moisture deeper into materials, and subsequent drying may not be complete. Weak points are invariably the mortar joints where pointing has deteriorated. Even minute cracks can draw much moisture deep into walls.

Patchlike staining on an interior wall may well be a sign of rain penetration. Rain rarely penetrates a thick, load-bearing masonry wall, however, and an owner should not accept recommendations from a company or contractor proposing to apply a water-repellent coating or carry out other remedial treatments without the advice of a competent and impartial independent expert.

Rain can cause problems at the base of a building if allowed to splash back from sidewalks, roadways and other hard surfaces adjacent to buildings or to form puddles against walls. This condition — called back-splash — can nearly saturate the walls and can be mistaken for ground-water problems on both the exterior and the interior — problems that are much more difficult to correct than back-splash (page 188).

Ground Water and Rising Damp

In some areas, ground-water levels are only a few feet below the surface. During heavy rains, the level of the subterranean water table can rise to the surface so that all the ground, from the surface down, is completely saturated. A broken water main, damaged storm pipe or clogged drain can create similar conditions.

Typical horizontal tidemarks. Efflorescence gradually diminishes above the line of crystallized salts. (Baird M. Smith)

Porous building materials act like a wick, drawing up moisture from damp ground. How effective this suction, or capillarity, is in drawing moisture in and up depends on the supply of moisture, the rate of evaporation and forces of gravity. This mechanism is known as rising damp and is generally identified by a horizontal stain or "tidemark" on the wall.

The tidemark is made up of large visible accumulations of crystallized salts, called efflorescence. Below the tidemark, the wall remains damp and the salt remains in solution. Above the tidemark, the wall remains relatively dry and therefore salt free. The band of salt crystals can be wide, because the tidemark can rise and fall as the equilibrium changes from one season to the next and as the ground-water level moves or the weather changes.

CONDENSATION

Condensation occurs when moist air reaches its dew point, that is, when it is cooled to the point at which water in the air can no longer be held as vapor. It is possible to avoid condensation by controlling air temperature or relative humidity.

On nonporous materials such as glass, condensation occurs on the surface in noticeable droplets. In porous materials,

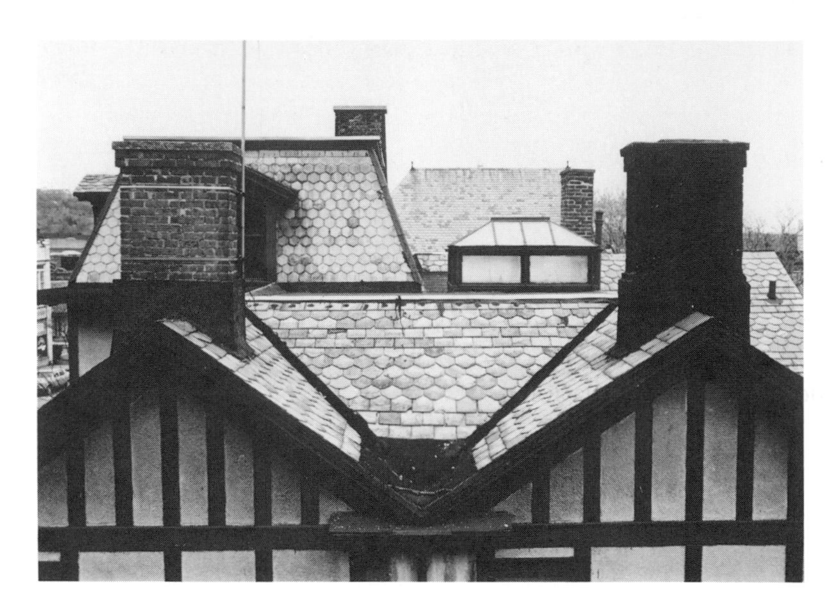

Rooftops. Drainage problems often begin here at the chimney, around flashing and near gutters. (Harriet Crane)

such as masonry, condensation occurs within the material. Under extreme conditions of low temperature, condensed moisture will freeze and can crack exterior brick or stone, causing the surface to spall.

Design and Maintenance Problems

Building design defects and poor or little maintenance are key causes of excessive moisture accumulation.

Design and construction errors

Inherent structural defects can be difficult to correct. Architects and builders may occasionally try to cut costs or save time by using poor-quality materials or using materials in ways in which they were not intended. Poorly designed roofs may not drain properly; parapets may lack proper cap flashing; improperly installed basement concrete slabs may fail to exclude damp.

Maintenance

Lack of maintenance can also cause or exacerbate moisture problems. More common than building defects, these maintenance-related problems are normally easy to remedy and may include clogged or broken rainwater pipes, leaky roofs, deteriorated flashing or loosely fitting window frames.

Improper maintenance operations leading to moisture problems include using excessive water to clean floors and planting flowers and shrubs adjacent to walls and watering them too often.

Earlier repairs or alterations

If improperly done, previous repairs also may have caused problems. For example, a weak masonry wall may be strengthened by injecting grout under pressure; but this grout can fill wall cavities meant to separate the exterior and interior walls and to provide ventilation.

The possibility of trapping moisture through use of an inappropriate surface treatment has been discussed on page 139. Changes to site landscaping, roof alterations or extensions, and blocking-in basement doors or windows with masonry can also cause serious moisture problems.

Obvious moisture problem. Water has been allowed to accumulate on this flat roof, perhaps because of a clogged pipe. The moisture has dripped behind the wall, damaging both masonry and wood building materials. (Paul Kennedy)

Damage to sandstone from freezing. Sandstones with very small pores are more susceptible to freezing. (Baird M. Smith)

Mechanical equipment

Furnaces, air conditioners and humidifiers can cause or exacerbate moisture problems if they are improperly used or malfunctioning. Poorly controlled humidifiers can lead to excessive humidity levels and serious condensation problems. Simple, inexpensive humidity gauges should always be a part of a humidification system, which should shut off whenever humidity rises above 35 percent in the winter.

WHAT MOISTURE DOES TO MASONRY

Building materials always have some moisture in them. This natural moisture content depends on the moisture content of the surrounding air and varies widely according to the type of material involved. The natural moisture content for some common materials is plaster, under 1 percent of its total volume; brick, about 1 percent; limestone, about 4 percent; and timber, up to 20 percent.

Once excessive moisture gets into masonry, the way it moves depends on the inherent properties of the building material, particularly its porosity and permeability.

Porosity is the ratio of the volume of pore space of a material to its total volume, that is, the percentage of its volume that is not solid. Light, soft brick with a porosity of 25 to 35 percent is much more likely to absorb moisture than dense, hard granite with a porosity of 1 to 2 percent.

Permeability refers to the extent to which the pores in a solid are interconnected and will, therefore, permit liquids to pass through. Brick is highly permeable because most of its pores are interconnected. Limestone also can be quite permeable, although its porosity is low — usually less than 15 percent. Permeability is measured in perms, units of measurement based on rates that vapor evaporates. The higher the perm value, the more likely a material will transmit moisture.

At the saturation point, every interconnected pore is filled with water and water can literally flow through the material. Saturation of masonry walls would cause considerable damage to mortar and excessive exterior and interior staining; few walls above ground ever reach this point, however, unless water pipes have burst or downspouts have clogged.

Left: Damage to brick from freezing moisture, most likely from back-splash. (Baird M. Smith)

Far left: Damage to granite from freezing. The tiny pore system of this stone, generally very hard and durable, makes it vulnerable to freezing. (Baird M. Smith)

Measuring moisture content on the surface and within materials (with handheld moisture meters or laboratory techniques) is invaluable in diagnosing the source of moisture. For example, higher moisture content at the surface would point to condensation as the source of the problem, whereas higher levels within the wall would suggest rising damp.

Moisture content in a structure does not fluctuate rapidly except at the surface. It can easily take several years for excessive moisture in a masonry wall to drop to the natural level even after a remedial treatment such as the installation of a damp-proof course.

Salts

Moisture present in building materials always carries salts; these come from the materials themselves (especially from mortars and most limestones), from air and rainwater, from salt-charged ground water and from sodium and calcium chloride used to melt ice on sidewalks and stairs.

When the water of a saline solution evaporates, it leaves behind salt crystals (efflorescence). Although this phenomenon normally does not create a problem when limited to the outermost surface, the pressures associated with this crys-

Detail of efflorescence. By identifying the type of salt, laboratory tests can assist in pinpointing the source of moisture. (Baird M. Smith)

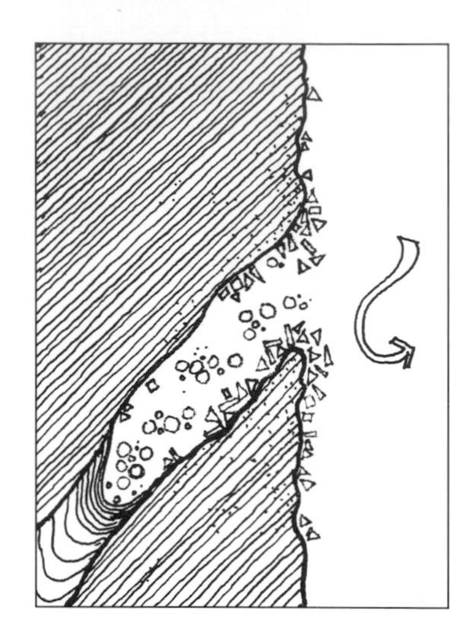

Efflorescence (salt crystals at the surface) and subflorescence (salt crystals within the material) in a stone or brick pore. As water evaporates, salt crystals form on both the exterior and the interior of the brick.

tallization can cause spalling when it occurs up to ½ inch from the surface of porous materials (subflorescence). Here, expansive forces can overcome the strength of the material, leaving the core largely unaffected, however.

Because moisture moves toward areas of higher salt concentration and evaporation, salty ground water will continuously move by capillary action toward wall surfaces, depositing fresh salt crystals at the points of evaporation and causing increased efflorescence and subflorescence over time.

Some of these salts, notably sodium chloride, absorb moisture and redissolve. It is not uncommon during periods of high humidity for highly salt-charged walls to absorb more moisture and become very damp. Because the moisture content of a wall with sodium chloride salts can easily be doubled or tripled during periods of high humidity, it can lead to inaccurate diagnoses of moisture problems. Inaccurate readings are also likely with moisture meters where there is heavy salt concentration. Meters operate by measuring electrical conductivity, which is higher in a salt solution and noticeably higher where salt has built up. Therefore, to ensure accuracy, readings should be made on several occasions during the year.

Damage to Materials

Common building materials differ markedly in their ability to resist damage from moisture. A level of moisture content that might cause slight staining or other tolerable minor damage in one material may produce severe staining, decay or spalling in another.

Brick is not necessarily damaged by the presence of moisture. In fact, most brick could remain submerged for decades and not be affected. Damage from subflorescence alone usually takes many years, and often decades, to occur. Cyclical changes such as freezing and thawing, however, do lead to damage when the expansive forces of subflorescence combine with the effects of freezing to exceed the strength of the brick. The result can be spalling or cracking almost overnight.

Sandstone, limestone, mortar, stucco and plaster, like brick, can be adversely affected by the expansive forces of subflorescence and freezing. And, like brick, these materials are not damaged by efflorescence beyond the resultant visual blemish.

Because these materials (with the exception of some varieties of sandstone) contain calcium, they are susceptible to damage from continued contact with water. Both calcium carbonate and calcium sulfate will be leached out of the material, leaving a weakened internal physical structure and often an intractable surface encrustation. Weaker materials, such as plaster and lime mortars, can be seriously damaged from this action. Durable stone, on the other hand, may require hundreds of years of continued moisture saturation before its strength is seriously weakened.

Treating Moisture Problems

The basic procedure for diagnosing masonry problems outlined earlier should be followed to identify the nature, extent and severity of moisture problems and then to diagnose their cause. Here are some additional steps to take in diagnosing and treating moisture problems.

When looking for background information, try to find out about past efforts in treating dampness problems, patterns of recent building use (especially the extent of heating, cooling or humidification) and characteristics of the local climate (temperatures, relative humidity, patterns of prevailing winds, the driving rain index during different seasons).

During the building inspection, identify and record evidence of dampness, note potential sources of unwanted moisture (for example, in bathrooms, kitchen, laundries) and verify the functioning of roofing, gutters, flashing, copings and downspouts, the most common source of moisture problems. Also, determine whether there is a damp-proof course. Physical probing may be necessary. Finally, compare the height of the ground floor relative to the earth outside and check out the site drainage patterns. Because excess moisture does not always leave visible evidence such as staining, salts or corrosion, a moisture meter should be used to measure high but invisible moisture levels.

With complex moisture problems, it may be necessary to undertake selected physical examinations, tests or long-term monitoring to differentiate among condensation, ground water or rain penetration as sources.

Condensation can be dismissed as a likely source if it can be ascertained, using a psychometric chart, that temperature and humidity conditions do not reach the dew point.

Rising damp from ground water can be eliminated if ground-water levels are well below the foundations. Consult with a city engineer, undertake soil borings or dig small test pits or 6-inch post holes to at least the depth of the base of the footings; if ground water is encountered or if water seepage fills the hole, then rising damp could be occurring.

If no ground water is encountered, the source of the moisture problem is probably limited to rainfall and attendant conditions such as poor site drainage, puddling at foundation walls or back-splash.

Rain penetration usually can be traced to junctions in the building such as flashing and windows. Penetration through the masonry walls themselves can be dismissed as a cause of moisture problems if there is a cavity in the walls or if the walls are more than 12 inches (three bricks) thick and in generally sound condition.

If it is unclear whether condensation is occurring, monitor, with a hygro-thermograph, the temperature and humidity levels over an extended period of time, preferably through a winter and summer.

Additional physical probings can help verify the exact wall configuration, the presence of a damp-proof course, foundation wall conditions, basement slab details and the condition of mortar, stucco or plaster. Detailed paint layer examination can help determine if coatings are acting as vapor retardants or breathable layers. Laboratory identification of efflorescent salts may be used to help identify sources of moisture.

Choosing Treatments

Be cautious in treating moisture problems in old and historic buildings. Obvious maintenance-related problems such as damaged roof flashing should receive immediate attention, but reckless intervention can do more harm than good. It can upset the sensitive balance between the environment and the building. The addition of excessive heat, extreme humidification or rapid drying damage old buildings. Proper diagnoses can take up to a year, and many years may be needed for treatments to work properly. Also, it is often good practice to address one problem at a time, monitoring its full effects before treating another one.

As stated earlier, the goal of intervening in old buildings is to protect and preserve them well into the future. With buildings of great historical significance, the first priority must be given to preserving building materials and contents. If there is a clash between building preservation and human comfort — for example, heating a previously unheated building, which could cause increased efflorescence — preserving historic building materials might be given precedence.

Even in other old buildings, treatments for moisture problems should be chosen to preserve the original materials as much as possible. In addition, old and historic buildings should not be the testing ground for untried materials or techniques. Up to 30 years of use have been necessary to demonstrate the competency of some materials, and many of the most modern and promising ones (sheet zinc, concrete, Knapen tubes, active and passive electro-osmotic systems) have failed to some degree.

If multiple treatments must be used, address treatments for rain and condensation problems first. Correcting these problems may allow the masonry to dry out enough to handle other moisture problems naturally. It should be kept in mind that successful treatment may not be possible for all moisture problems. Treatment may be too expensive or impractical, as in a building with very thick or irregular stone walls where severe rising damp may not be fully treatable. Drastic measures, such as rebuilding walls or foundations or wholesale demolition, is not recommended for buildings of historical significance. Excessive moisture can be reduced and some damage repaired, but arresting the problem may be beyond the capabilities of the present technology. The proper treatment for intractable moisture problems may be to change the building's use. For instance, a building may not tolerate the humidity necessary for a museum but may function well as office space.

Finally, it may be better to live with minor damage, such as an odd stain or a patch of efflorescence, and to clean or repaint periodically, rather than go to the trouble and expense of major repairs.

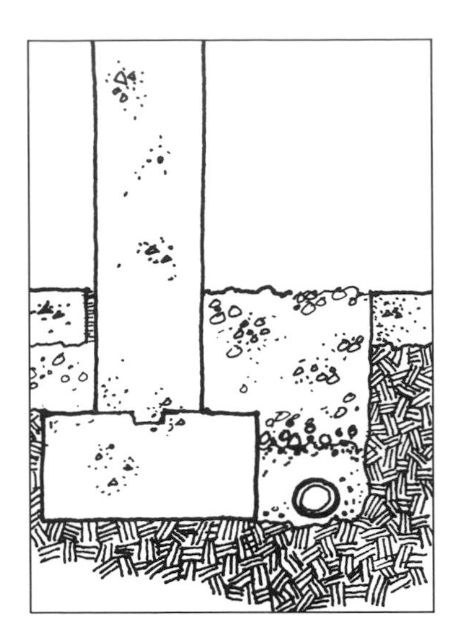

French (footing) drain. A perforated drain pipe is placed next to the footing to drain both ground water and excess rainwater. The drain should rest on a bed of gravel and be covered with at least 6 inches of gravel.

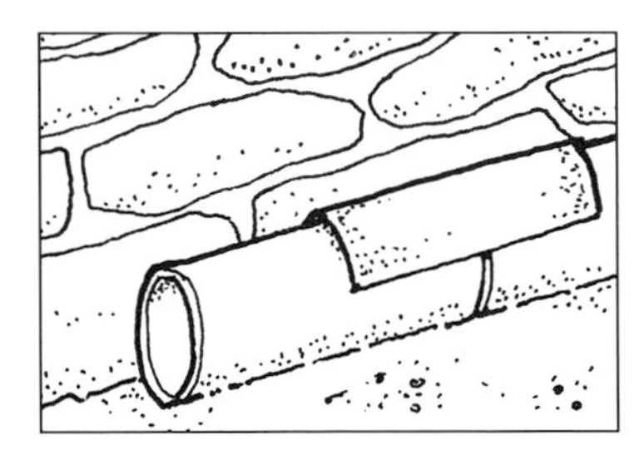

Terra cotta weeping pipe. A gap of about 1/4 inch, left between sections, is covered with tar paper to prevent earth and sand from clogging the drain.

Improving Drainage

The simplest and most important method to improve drainage (that is, to ensure that rainwater is directed away) is to slope the ground away from the building and extend the roof downspout farther out.

Another common treatment for reducing, but not necessarily eliminating, rising damp and related moisture problems arising from ground water and poor site drainage is to install footing drains, also called French drains. These drains reduce the quantity of water affecting the wall by collecting rainwater falling near the building and by lowering high ground-water levels. They also collect the underground water and evacuate it to a drainage well far from the building or connected to a nearby storm sewer. The slope given the pipes ensures proper drainage.

French drains are traditionally installed beside the base of the foundations by laying short sections of perforated terra cotta or concrete pipes end to end, slightly spaced apart. A perforated plastic (PVC) weeping pipe also can be installed. The holes in any of these materials collect the water but can also collect sand or earth, clogging up the system. Roots seeking water also can make their way into the drain. As a preventive measure, the drain, or at least the gaps between the sections of the drain, are covered with building paper or another filtering material.

Footing drains are most practical where the entire periphery of the building is readily accessible and water can be drained away from the building without pumping. Urban sites pose several complications: inaccessibility of the building periphery to trenching; inability to estimate the quantity of ground water; difficulties in establishing property boundaries and limiting liability for damage to adjacent properties; and inability to establish positive drainage of collected water, so that pumps are required.

Two notes of caution:

- Building foundations may be disturbed by the digging of a large deep trench, and buildings on wooden pilings or footings can be severely jeopardized when ground-water conditions are altered. An expert such as a hydraulic or civil engineer should be consulted before installing drains.

■ Any disturbance of the ground near historic buildings should be accompanied by protection and identification of archeological resources.

Because the cost of footing drain systems can be low, they may be a first choice, especially if the site does not drain well. Drains cannot guarantee the arrest of rising damp, but even a slight change in the soil's moisture equilibrium could reduce the problem substantially. After installation, the effect should be monitored by taking moisture readings of the masonry.

Waterproofing Foundations

Foundation walls may be protected from water penetration by applying on the exterior or interior one of a few techniques available.

Excavation of a foundation in preparation for waterproofing. (© Richard Pieper)

On the outside

Traditionally, foundation walls were waterproofed with hydraulic lime pargings; these materials were replaced in the latter part of the 19th century with hot applied bitumen products such as tar or asphalt, which have proven less than completely effective on irregular rubble foundations where mortar joints are often quite deteriorated.

It is best first to parge the wall to provide an even surface to the foundations on which the waterproofing can be applied.

When infiltration problems are more serious, a bitumen multilayered membrane, similar to that used for flat roofs, can be installed. However, it is often quite difficult to obtain a perfectly waterproofed barrier when installing such membranes on a foundation.

Clay also can be used to waterproof foundations. It is often employed as a backfill to solve problems of water drainage in the soil or on the surface. Clay is waterproof once it is saturated with water. Because it does retain water, however, it may freeze and cause the foundation to heave. Clay can be pumped to the site and is also available in the form of a powder or in prefabricated molded panels, which are nailed to the foundations. These panels become watertight when in contact with the water present in the ground.

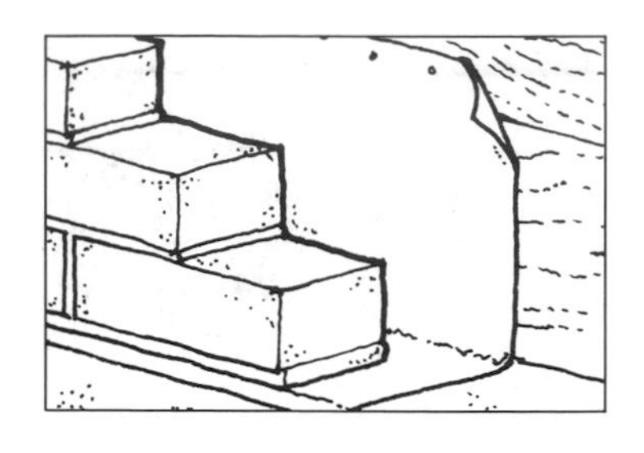

Flashing in masonry veneer construction acting as a waterproof barrier.

On the inside

Methods applied on the interior are not very effective; they do not stop water from penetrating into the masonry.

For stone foundations, water penetration through joints may be lessened by repointing the deteriorated joints. The harmful effects of salt can be reduced by plastering the wall with ¾-inch-thick parging, which works as a sacrificial layer and must be periodically replaced.

If water is infiltrating through the foundation wall and it is impossible to waterproof from the outside, it is best to provide a path where the water can be directed without harming the building. This involves installing the interior finish (plasterboard or flooring) on furring strips to allow the water to find its way down to floor drains. The cavity also maintains good air circulation behind the finishes to help keep them dry.

INSTALLING DAMP-PROOF COURSES

An excellent treatment for rising damp is the installation of a damp-proof course, a barrier in the lower part of the wall that prevents moisture from rising vertically. The method is commonly used in Britain and Europe, but, unfortunately, its use in North America is quite limited; many European materials are not even available here.

In traditional construction, the foundation was sometimes made of a less porous stone such as granite; alternatively, copper flashing was installed on top of the foundation and below the masonry.

Once a damp-proof course is installed, the wall area below it continues to suffer subflorescence or freeze and thaw problems and should be maintained continually and repaired every 10 to 20 years.

Physical damp-proof courses

An impermeable material, such as slate, which is chemically inert and has relatively little potential for decay, may be installed in the wall to create a physical damp-proof course. Some barriers have been in service for more than 100 years, and new ones can be expected to last the life of a building.

Physical damp-proof courses should be strongly considered, as they will arrest rising damp for most ordinary building types. Where installation is practical, or possible, they have the best all-around performance; durability and the extra cost should be acceptable when the long-term preservation benefits are considered. Damp-proof course materials are included under flashing or waterproofing materials in architectural product listings.

Roofing slates as a damp-proof course can be installed in an existing wall by removing two courses of brick, or one course of stone, and inserting a double layer, lapping one over the other, through the entire thickness of the wall. Generally slates are installed in widths of about 3 linear feet, while the above masonry is braced. Installers work in alternate areas along the length of the wall. Some building settlement can result, but modern nonshrinking mortars almost eliminate the problem.

Another installation technique for a physical damp-proof course involves inserting flashing, a thin, impermeable material, into a cut-out mortar joint. The flashing consists of semirigid sheets of bituminized felt or impregnated fiberglass cloth. Copper, lead or aluminum sheets coated with tar or rubber are used on multistory buildings with great compressive loads.

In Europe effective techniques have been developed to cut out mortar joints using a high-speed carborundum mason's saw or an adapted chain saw, which, while inappropriate for repointing in general, may be acceptable for installing a damp-proof course. Handsawing and removing brick by hand are less likely to damage the masonry but more time consuming and expensive. These systems generally can be installed by skilled builders or masons, but it is almost inevitable that

Top: Dismantled brick wall showing a slate damp-proof course. (Baird M. Smith)

Above: Cavity wall with newly exposed tar damp-proof course. (Baird M. Smith)

Injected chemical damp-proof course installed above another type of damp-proofing — two courses of high-fired brick. (Baird M. Smith)

some damage will occur to the masonry units as a result of the saw blades, and all techniques create some dust and inconvenience. Installation for residential-scale jobs can be completed in a few days.

Chemical damp-proof courses

Until recently, chemical damp-proof courses, an effective and low-cost protection against rising damp, have not had much use in North America, despite their wide acceptance in Britain and Europe. The method involves drilling holes in the masonry wall and injecting agents that solidify into a hard, continuous waterproof layer running the full thickness of the wall or feeding in low molecular-weight silicones that create a zone of water repellency within the wall.

Passive electro-osmotic system. The thickened mortar joint holds the copper strand, which is grounded to the earth. (Baird M. Smith)

Passive electro-osmotic systems

In passive electro-osmotic damp-proofing systems, a grounded copper strand inserted into a horizontal joint, purportedly, sets up a zone with a different electrical potential from the rest of the wall, thus stopping the movement of moisture; such systems have not proven effective, however, and cannot be recommended. Active systems, involving use of a continuous electrical current, do work but are subject to rapid corrosion and, therefore, are not an appropriate long-term solution.

Limiting Rain Penetration

Rain penetration is not normally a problem in old masonry building and is unlikely in cavity walls and solid walls more than one foot thick. In those rare cases where true evidence of interior staining caused by severe rain penetration exists and exterior flashing at the parapet, windows, doors or other joints has been repaired, then the problem is almost certainly the condition of the mortar joints. Only if rain penetration continues, after the joints have been repointed, should more drastic measures be considered — and only where the penetration is so severe that staining or damage still occurs.

Surface treatments should be considered only in limited cases, because their use is rarely successful without undesirable side effects; the various surface treatments discussed in Repairing Deteriorated Surfaces and Building Parts may be consid-

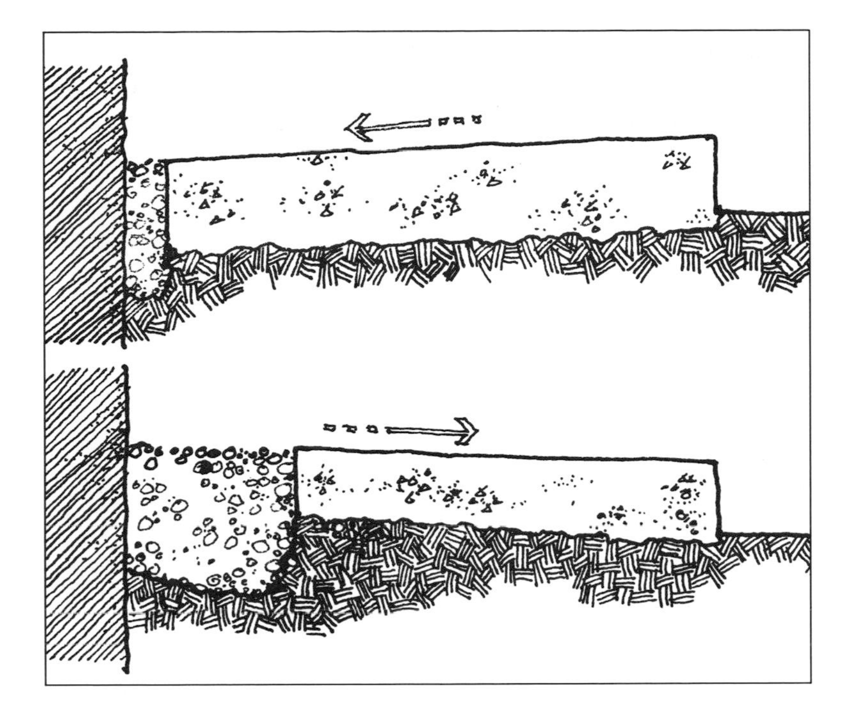

Treatments for rain back-splash and puddling. An area of pavement next to the building is cut out and filled with gravel (top), permitting water to drain through the gravel — a treatment possible with existing pavement. The preferred treatment (bottom) is to cut out a large area, fill it with gravel and slope the pavement slightly away from the building.

ered after careful evaluation and testing by an independent professional.

Rain back-splash and puddling should be avoided by ensuring that roof runoff is controlled with proper gutters and downspouts and that the ground drains away from the building. Paving materials at the base of walls should be at least 2 inches from the building. A gravel strip 2 or more feet wide reduces back-splash and helps drain away water.

Treating Salt Encrustations

After successful treatment of a primary moisture problem, such as rising damp or rain penetration, walls begin to dry, and efflorescence and subflorescence are the unavoidable results. The key to treating these salts is to ensure that the crystallization takes place in a controlled way (by using parging as a sacrificial layer or by periodic cleaning) without causing further damage.

In severely salt-encrusted walls, a damp clay poultice (page 105) can be applied to draw moisture and salts from the wall, repeating this as necessary until salt content is sufficiently reduced to permit redecorating on the interior.

One

Further Reading

The following is a list of basic publications for those looking for more detailed information. Several of these publications served as the main source material for this book. For more complete bibliographies, refer to the documents listed here, particularly the *Masonry Conservation and Cleaning Handbook* of the Association for Preservation Technology.

Tor House (1919–24), Carmel, Calif., constructed by poet Robinson Jeffers of granite stones brought up from the nearby beach. (John Frisbee)

Opposite: Elfreth's Alley (1703), Philadelphia, a narrow alley lined with simple brick townhouses. (HABS)

Association for Preservation Technology

This association is an international organization for preservation professionals. An annual conference in the United States or Canada, a newsletter, *Communiqué,* and a quarterly bulletin, which includes articles on state-of-the-art restoration techniques, are part of the organization's yearly agenda. To join or order any of the publications listed below, write APT, P.O. Box 2487, Station D, Ottawa, Ontario K1P5W6, Canada.

APT Bulletin 9, no. 2 (1979), "The Chemistry of Masonry Building."

APT Bulletin 17, no. 2 (1985), special issue on masonry.

Masonry Conservation and Cleaning Handbook. Keith Blades, Gail Sussman and Martin Weaver, eds. 1984. Architects, contractors and other professionals who deal with old masonry should have their own copy of this excellent collection of 27 articles and other documents. Prepared for a professional training course, it emphasizes techniques and applications. For example, there are seven articles on cleaning techniques that include detailed descriptions of the methods discussed in this book, including specific chemical formulations for use in removing various types of stain. The handbook contains, besides an extensive bibliography, a 44-page annotated Master Specification for the Cleaning and Repointing of Historic Masonry prepared by the Ontario Ministry of Citizenship and Culture.

Illinois Historic Preservation Agency

Two leaflets have been published by the agency. They provide additional useful information on stucco and concrete. Write to Publications Editor, Illinois Historic Preservation Agency, Old State Capitol, Springfield, Ill. 62701.

Illinois Preservation Series no. 2. *Stucco.* Brian D. Conway and Richard S. Taylor, eds. 1980.

Illinois Preservation Series no. 8. *Concrete in Illinois: Its History and Preservation.* William B. Coney and Barbara M. Posadas. 1987.

National Academy Press,

National Academy of Sciences

This organization has published a useful compilation of 21 papers focusing on chemical, physical and engineering analyses and treatments for old buildings. Of interest primarily to professionals, the publication was initially prepared by the National Materials Advisory Board Commission on Engineering and Technical Systems of the National Research Council. Order from the National Academy Press, 2101 Constitution Avenue, N.W., Washington, D.C. 20418.

Conservation of Historic Stone Buildings and Monuments. National Research Council. 1982.

National Park Service, U.S. Department of the Interior

The Technical Preservation Services branch of the National Park Service publishes a number of helpful publications of interest to professionals and laypersons alike. Of particular note are a series of publications called Preservation Briefs. Those that deal with masonry issues are listed here, along with three other publications. For a publications catalog, write Technical Preservation Services, Preservation Assistance Division, National Park Service, P.O. Box 37127, Washington, D.C. 20013-7127.

Preservation Brief no. 1. *The Cleaning and Waterproof Coating of Masonry Buildings.* Robert C. Mack, AIA. 1975.

Preservation Brief no. 2. *Repointing Mortar Joints in Historic Brick Buildings.* Robert C. Mack, de Teel Patterson Tiller and James S. Askins. 1980.

Preservation Brief no. 5. *Preservation of Historic Adobe Buildings.* de Teel Patterson Tiller and David W. Look, AIA. 1978.

Preservation Brief no. 6. *Dangers of Abrasive Cleaning to Historic Buildings.* Anne E. Grimmer. 1979.

Preservation Brief no. 7. *The Preservation of Historic Glazed Architectural Terra-Cotta.* de Teel Patterson Tiller. 1979.

A Glossary of Historic Masonry Deterioration Problems and Preservation Treatments. Anne E. Grimmer. 1984.

Moisture Problems in Historic Masonry Walls: Diagnosis and Treatment. Baird M. Smith, AIA. 1984.

New York Landmarks Conservancy

This organization has published a manual on the rehabilitation of historic building facades and a leaflet on sandstone restoration. Write to Publishing Center for Cultural Resources, 625 Broadway, New York, N.Y. 10012-2662.

Historic Building Facades: A Manual for Inspection and Rehabilitation. Robert E. Meadows. 1986.

The Maintenance and Repair of Architectural Sandstone. Michael Lynch and William Higgins. 1982.

THE OLD-HOUSE JOURNAL CORPORATION

This organization publishes *The Old-House Journal,* a bimonthly magazine. While full of practical advice for the do-it-yourselfer, OHJ is also of interest to professionals. Each issue features several articles on specific rehabilitation problems as well as sources of information and restoration products. A year's back issues are reprinted as *The Old-House Journal Yearbook.* In addition, the annual *Old-House Journal Catalog* lists more than a thousand companies that supply products and services for restoration. Some specific issues dealing with masonry are listed below. Contact the Old-House Journal Corporation, 69A Seventh Avenue, Brooklyn, N.Y. 11217.

The Old-House Journal (July 1982), "Patching Limestone and Marble." Lynette Strangstad.

The Old-House Journal (August 1982), "Patching Brownstone." Lynette Strangstad.

The Old-House Journal (January 1987), "Special Masonry Report." Susan M. Tindal.

THE PRESERVATION PRESS,

NATIONAL TRUST FOR HISTORIC PRESERVATION

The publisher of the present book has three other titles of special interest to those studying a particular masonry problem. Order from the Mail-Order Division, National Trust for Historic Preservation, 1600 H Street, N.W., Washington, D.C. 20006.

All About Old Buildings. Diane Maddex, ed. 1985.
This book includes many useful references that can help in planning a masonry rehabilitation project. Included also are lists of general rehabilitation manuals, state historic preservation offices and university programs in historic preservation.

Introduction to Early American Masonry: Stone, Brick, Mortar and Plaster. Harley McKee. 1973. This is an excellent book on traditional masonry uses, styles and construction techniques in the United States.

Respectful Rehabilitation: Answers to Your Questions About Old Buildings. National Park Service. 1982. This guide answers 150 of the most-asked questions, many dealing with masonry, about rehabilitating old and historic buildings.

Regional Offices,

National Trust for Historic Preservation

The National Trust maintains six regional offices and one field office that provide advice and consultation to individuals and organizations seeking assistance with a wide range of preservation problems from referrals to financial assistance. Contact the office within your region.

Northeast Regional Office
45 School Street, Fourth Floor
Boston, Mass. 02108

Mid-Atlantic Regional Office
6401 Germantown Avenue
Philadelphia, Pa. 19144

Southern Regional Office
456 King Street
Charleston, S.C. 29403

Midwest Regional Office
53 West Jackson Boulevard, Suite 1135
Chicago, Ill. 60604

Mountains/Plains Regional Office
511 16th Street, Suite 700
Denver, Colo. 80202

Texas/New Mexico Field Office
500 Main Street, Suite 606
Fort Worth, Tex. 76102

Western Regional Office
One Sutter Street, Suite 707
San Francisco, Calif. 94104

Index

Page numbers in italics refer to illustrations and captions.

AUTHOR

Mark London is an architect working in Montreal, Quebec, Canada. Currently a planner for the city of Montreal, he directed the activities of the preservation organization Heritage Montreal from 1980 to 1987. As an architect in private practice, London was instrumental in the planning of many mixed-use development projects and handled numerous smaller projects involving rehabilitation of houses and stores, an apartment building and a library. He also has conducted extensive building inspections. As an educator, London has developed programs in architectural conservation and rehabilitation for the University of Montreal and was one of the founders of a successful series of workshops on home renovation in Montreal. He writes regularly on building rehabilitation for newspapers and magazines.

Gargoyle from the National Cathedral, Washington, D.C. (Stewart Brothers)

F Street portico of the Patent Office (1867), now the National Portrait Gallery and the National Museum of American Art. (Jack E. Boucher, HABS)

REHABILITATION

Houses by Mail: A Guide to Houses from Sears, Roebuck and Company

Katherine Cole Stevenson and H. Ward Jandl. A unique history and guide to nearly 450 precut house models sold by Sears from 1908 to 1940. 408 pp., illus., biblio., index. $24.95 pb.

Introduction to Early American Masonry: Stone, Brick, Mortar and Plaster

Harley J. McKee, FAIA. A classic guide to masonry construction that examines the origins, use, manufacture and styles of each masonry type. 92 pp., illus., biblio., index. $9.95 pb.

New Energy from Old Buildings

Details the energy conservation benefits of old buildings and how to safeguard them during retrofitting. 208 pp., illus., gloss., biblio., index. $9.95 pb.

Respectful Rehabilitation: Answers to Your Questions About Old Buildings

National Park Service. A "Dear Abby" for old buildings that answers 150 of the most-asked questions about rehabilitating old houses and other historic buildings. 200 pp., illus., biblio., index. $12.95 pb.

Building Watchers Series

America's Architectural Roots: Ethnic Groups That Built America

Dell Upton, Editor. Highlights 22 ethnic groups, from Africans to Ukrainians, that have shaped the diversity of American architecture. 196 pp., illus., biblio., index. $9.95 pb.

The Buildings of Main Street: A Guide to American Commercial Architecture

Richard Longstreth. A fresh look at architecture found along America's Main Streets. Building types are documented with numerous illustrations from towns and cities nationwide. 156 pp., illus., biblio., index. $8.95 pb.

Built in the U.S.A.: American Buildings from Airports to Zoos

Diane Maddex, Editor. A guidebook-size history of 42 American building types, showing how their forms developed from their functions. 192 pp., illus., biblio., $8.95 pb.

Master Builders: A Guide to Famous American Architects

Introduction by Roger K. Lewis. Profiles of 40 major architects who have left indelible marks on American architecture — from Bulfinch to Venturi. 204 pp., illus., biblio., index. $9.95 pb.

What Style Is It? A Guide to American Architecture

John Poppeliers, S. Allen Chambers, Jr., and Nancy B. Schwartz, Historic American Buildings Survey. One of the most popular, concise books on American architectural styles. A portable guidebook designed for easy identification of 22 styles of buildings. 112 pp., illus., biblio., gloss. $7.95 pb.

Historic Interiors Series

Fabrics for Historic Buildings

Jane C. Nylander. 3rd edition. A popular primer that gives practical advice on selecting and using reproductions of historic fabrics. Includes an illustrated catalog listing 550 reproduction fabrics and a list of manufacturers. 160 pp., illus., gloss. biblio. $12.95 pb.

Floor Coverings for Historic Buildings

Helene Von Rosenstiel and Gail Caskey Winkler. Presents historical overviews of authentic period reproduction floor coverings from wood floors and linoleum to hooked and woven rugs. Also provides purchasing information for 475 patterns available today. 284 pp., illus., gloss., biblio. $12.95 pb.

Lighting for Historic Buildings

Roger W. Moss. Describes the history of lighting from the colonial era through the 1920s and arranges several hundred reproduction fixtures by original energy source (candle, gas, gas-electric and electric) and type (chandeliers, sconces and lanterns). 200 pp., illus., gloss., biblio. $12.95 pb.

Wallpapers for Historic Buildings

Richard C. Nylander. Shows how to select authentic reproductions of historic wallpapers and where to buy more than 350 recommended patterns. 128 pp., illus., gloss., biblio. $12.95 pb.

Great American Places Series

Great American Bridges and Dams

Donald C. Jackson. The first state-by-state guide to 330 of the best-known American bridges and dams. Provides a historical overview and discusses the preservation of these endangered structures with engineering and architectural information on each entry. 360 pp., illus., biblio., index. $16.95 pb.

Great American Movie Theaters

David Naylor. The first state-by-state guide to 360 of the most dazzling and historic movie theaters still standing. Gives colorful architectural and historical descriptions of these magnificent landmarks. 276 pp., illus., biblio., index. $16.95 pb.

To order, send the total of the book prices (less 10 percent discount for National Trust members), plus $3 postage and handling, to: Mail Order, National Trust for Historic Preservation, 1600 H Street, N.W., Washington, D.C. 20006. Residents of Calif., Colo., Wash., D.C., Ill., Iowa, La., Md., Mass., N.Y., Pa., S.C., Tex. and Va. please add sales tax. Make checks payable to the National Trust or provide your credit card number, expiration date, signature and telephone number.